Pathways for Getting to Better Water Quality: The Citizen Effect

T0135149

Lois Wright Morton · Susan S. Brown
Editors

Pathways for Getting to Better Water Quality: The Citizen Effect

 Springer

Editors
Lois Wright Morton
Iowa State University
Department of Sociology
Ames, IA
USA
lwmorton@iastate.edu

Susan S. Brown
Iowa State University
Ames, IA
USA
ssbrown@iastate.edu

ISBN 978-1-4899-8128-8   ISBN 978-1-4419-7282-8 (eBook)
DOI 10.1007/978-1-4419-7282-8
Springer Dordrecht Heidelberg London New York

Printed on acid-free paper

Springer is part of Springer Science+Business Media (www.springer.com)

*To a colleague and friend, a soil scientist
who inspired and encouraged*

Foreword

This book is about accomplishing change in how land is managed in agricultural watersheds. Wide-ranging case studies repeatedly document that plans, policies, and regulations are not adequate substitutes for the empowerment of people. Ultimately, change on the land is managed and accomplished by the people who live on land within each watershed. Change is also more rapid and sustainable in small watersheds, Hydrologic Unit Code (HUC) 12 or smaller, where people have common ties to their communities as well as to the land and water.

The 1972 Clean Water Act with amendments was the first modern legislation in the United States to focus on nonpoint source pollution from agricultural lands. Although the 1972 Act exempted agricultural land from permit requirements, the Act did provide USDA (United States Department of Agriculture) authority to conduct studies and provide technical and financial assistance to agriculture land managers. Subsequently, Congress provided funding for USDA to initiate the Rural Clean Water Program (RCWP). The implementation of RCWP involved 21 watersheds across the nation including Iowa, Kansas, and Nebraska which became 3 of the 4 states representing the Heartland Regional Water Coordination Initiative. Many of these watersheds were HUC 12 or smaller in size. The program focused on water monitoring, technical assistance for existing conservation practices, and the introduction of cost-share for improved and alternative BMPs (Best Management Practices). The RCWP summary report states that one-on-one contact by agency staff was the most effective way to ensure participation in the program (Gale et al. 1993).

The next major watershed program for agricultural lands that included educational, technical, and financial assistance was initiated by then President George Bush in 1990 and implemented by USDA in cooperation with other federal agencies. From 1990 through 1998, more than 70 Hydrologic Unit Area watersheds and 16 Demonstration Projects were funded through a 3-way partnership among USDA agencies: Cooperative States Research, Education, and Extension Service (CSREES), Farm Service Agency (FSA), and Natural Resources Conservation Service (NRCS). In 1998, Congress passed the Agricultural Research, Extension, and Education Reform Act (AREERA) which set the stage for competitive funding within CSREES (renamed the National Institute for Food and Agriculture, NIFA, effective October 2009). Within the guidelines of competitive funding, visionary program leaders at NIFA established the National Water Program which funded grants starting in FY00 and has commitments to fund some projects, including regional projects, through FY12. Therefore, NIFA, through its integrated research, education, and extension programs in partnership with the state land grant universities and colleges, has national responsibility for providing leadership for watershed research and extension education. Projected population growth and additional demands on the nation's water resources suggest that many challenges will remain to be addressed well beyond FY12.

I share this short history of watershed-based water quality projects to emphasize that environmental conservation work by extension educators has its roots in soil conservation and soil management. Educators had primarily worked with farmers on a field-by-field and farm-by-farm basis. The 1972 Clean Water Act with amendments provided the transition for extension educators to focus on small watersheds and the collective group of producers and associated communities within each watershed. Over time, the focus moved from the physical aspects of the watershed to include the human dimension and community attributes. Extension educators understand that changing practices implemented on the land is less about technology, tools, and technical plans than it is about people – what they think, what they value, and the choices they make. Management is not just science and specifications. It is also what people do. And within watersheds, bringing people together in a collective setting to focus on a common issue, water quality and the quality of the physical attributes of the watershed, reduces the one-on-one competitiveness, and sometimes friction, between neighbors. This effort can forge a group with shared leadership, goals, and the ability to create change.

I have had the opportunity to be involved in many different watershed projects starting in 1984. Experience suggests that there are three major components that must be addressed in the planning, development, and implementation of agricultural watershed projects. First and foremost, successful projects require a team approach consisting of individuals representing multiple disciplines. Understandably, extension educators are required to be technically competent in their respective subject matter expertise whether it be agricultural engineering, agronomy, animal ecology, economics, forestry, hydrology, soil science, or related disciplines. However, we have also found that without inclusion of social scientists, specifically rural sociologists, it is more difficult to document successful project outcomes. Second, designing a thorough evaluation approach for the project prior to implementation is essential. Benchmarks must be established in order to document changes in people's behavior as well as practices implemented on the land.

Third, an experienced and credible individual who is local to the area of the watershed project is a must for the on-site project coordinator. The real heroes of successful watershed projects are the coordinators on the ground working shoulder to shoulder with the people who manage the land and depend on the land for their livelihood. The coordinator must be skilled in more than technology, tools, and technical practices. Too often the on-site watershed project coordinator is an inexperienced individual lacking both human dimension skills and the confidence of the watershed citizens, both of which are critical for accomplishing successful outcomes. The most successful watershed project coordinators are comfortable and skilled in involving citizens in identifying project outputs and outcomes and sharing leadership for these decisions with local participants. Readers will find case studies in the following chapters that reinforce these observations.

The Citizen Effect is about citizens being proactive and setting the direction for the future of their watershed. It is about people and how they work together to manage improvements within their watershed. And it is about how educators and other public sector specialists can help them do so. Empowered citizenry will ensure there is a positive future for their children, their community, and the nation's water resources.

January 20, 2010   Gerald A. Miller
College of Agriculture and Life Sciences
Agriculture and Natural Resources Extension
Iowa State University
Ames, IA, USA

Reference

Gale JA, Line DE, Osmond DL, Coffey SW, Spooner J, Arnold JA, Hoban TJ, Wimberley RC (1993) Evaluation of the Experimental Rural Clean Water Program. U.S. Environmental Protection Agency (EPA-841-R-93-005). On-line http://www.water.ncsu.edu/watershedss/info/rcwp/ Verified January 17, 2010

Acknowledgement

This book was made possible through the Heartland Regional Water Coordination Initiative which is funded by the United States Department of Agriculture (USDA) National Institute for Food and Agriculture (NIFA) Integrated Water Program under agreements 2004-51130-002249 and 2008-51130-19526. Any opinions, findings, conclusions, or recommendations expressed in this publication are those of the author(s) and do not necessarily reflect the view of the US Department of Agriculture. Special thanks to the scientists and staff at USDA-Agricultural Research Service North Appalachian Experimental Watershed, Coshocton, Ohio for sharing space and resources. Thanks to LaDonna Osborn, Iowa State University, Department of Sociology for manuscript preparation. Thanks also to all those citizen-scientist farmers who shared with us their experiences and knowledge.

About the Authors

Charles Barden is a professor of forestry at Kansas State University, having joined the faculty in 1998. He frequently advises American Indian tribes and other entities in Kansas on natural resource issues. He was recently recognized as a Fellow of the Society of American Foresters. Barden earned his Ph.D. from Penn State, and his M.S. from Virginia Tech, both in Forestry. A native of Rhode Island, he earned his B.S. degree in Natural Resource Management from the University of Rhode Island. He enjoys conducting applied research, using trees to solve environmental problems (i.e., riparian buffers, phytoremediation, bioenergy, and windbreaks).

Terrie A. Becerra received her PhD in Sociology at Kansas State University in 2010 and is currently a post doctoral fellow at Oklahoma State University, Stillwater, OK. Her research interests are in the areas of environmental and natural resource studies, particularly regarding water resource issues, but also in sustainable agriculture and sustainable rural communities. Her publications include a coauthored article relating to water quality BMPs in Kansas. As a graduate student, she has been involved with watershed research concerned with the producer's adoption of agricultural conservation practices and their effect on water quality. Her dissertation research examines water governance structures in Kansas.

Kathryn Brasier is an Associate Professor of Rural Sociology at The Pennsylvania State University. Her research and extension programs focus on collective action and networking around agricultural and environmental issues. Specific interests include network effects on learning and innovation, particularly among farmers using conservation practices; the development and activities of community environmental organizations; civic engagement in local land use planning; the effects of space and scale on farm management and environmental decision-making; and spatial data analysis techniques and their uses in the social sciences. Dr. Brasier received her PhD in Sociology from the University of Wisconsin – Madison in 2002.

Susan S. Brown received her PhD in plant science from Virginia Polytechnic Institute and State University in 1980. In 1990, Dr. Brown became program specialist for water quality with Iowa State University Agronomy Extension. In that position she has coordinated planning and project development for more than 30 watershed and water quality projects and been co-principal investigator on 12 multi-funded projects involving extension, state, and regional natural resource agencies and organizations. Since 2004, Dr. Brown has served as Regional Liaison as well as a member of the Community Involvement and Human Dimensions issue teams for the Heartland Regional Water Coordination Initiative, a multistate project supported by the USDA NIFA National Integrated Water Quality Program.

Robert Broz is the Water Quality Specialist with the University of Missouri Extension – Water Quality Program. Bob has been with the University since 1993 and holds an M.S. in Extension Education and a B.S. in Animal Husbandry. Through his work with the Missouri Department of Agriculture and teaching in a public school system, he had extensive training in community development, facilitation, and group process.

Kristen Marie Corey is a graduate of Iowa State University with an M.S. in Rural Sociology and Sustainable Agriculture. For 2 years, she worked as a graduate research assistant for the Heartland Regional Water Coordination Initiative Project, coordinated by Iowa State University, the University of Nebraska-Lincoln, the University of Missouri and Kansas State University. Her work for this project included a master's thesis on the effect of the Endangered Species Act on a watershed in this four-state area. She presently works as a technical writer for the Iowa Department of Human Services in the policy unit.

Dan Downing is an Extension Associate with the University of Missouri Extension – Water Quality Program. He holds an M.S. in Extension Education, B.S. in Agriculture, and has extensive training in community development, group process, and group dynamics. He specializes in community coalition building around the issue of watershed management as well as general responsibilities relating to water and environmental issues. While working in over twenty watersheds, along with classroom instruction in the Food & Biological Engineering unit, he has authored such publications including: Helpful Hints for Working with Volunteers, Agency Roles in Watershed Management, and Watershed Committee Organizational Structures & Styles.

Ryan Dyer is the currently elected Tribal Treasurer of the Prairie Band Potawatomi Nation, a resident Tribe of Kansas. He continues to carry on a family tradition of service to Indian people. Dyer holds an A.S. in Natural Science from Haskell Indian Nations University and a B.S. in Mathematics from the University of Kansas. He participated in the Radio Ice Cerenkov Experiment as a research assistant while attending the University of Kansas. Dyer enjoys the company of his wife and two children, playing and coaching soccer, reading, and flying airplanes.

Lillian Fisher is a researcher at Haskell Indian Nations University (HINU). A native of New York, she earned a B.S. in Civil Engineering from Union College, Schenectady, NY and worked as a structural engineer in NYC for 3 years before enrolling in graduate school in biology. She earned an M.S. in Biology from Northeastern University, Boston, MA in 1996, and an A.M. in Biology from Harvard University, Cambridge, MA, in 1998. Fisher moved to Lawrence, KS and took her present position at Haskell in 2005, where she is charged with promoting undergraduate research, particularly in watershed and native plant sustainability.

Kristin Floress is an Assistant Professor of Human Dimensions in the Center for Land Use Education at the University of Wisconsin-Stevens Point. She teaches courses in integrated resource management, communications, and policy and her extension work involves managing an online master's degree program for working natural resource professionals. Her research focuses on collaboration and private land management.

Matt Helmers is an Associate Professor and Extension Agricultural Engineer in the Department of Agricultural and Biosystems Engineering at Iowa State University. Dr. Helmers is a native Iowan and received his PhD from the University of Nebraska-Lincoln. His research and extension focus at Iowa State is in the areas of water quality and water resources management. In particular, he is studying water quality effects of agricultural best management practices including strategic placement and design of buffer systems and methods to improve water quality in tile drained landscapes.

Zhihua Hu is a PhD candidate in sociology at Iowa State University. Her research interests include natural resources and the environment, quantitative methods, and community development.

Brian Lee is an Associate Professor in the Department of Landscape Architecture at the University of Kentucky. He applies geospatially based analyses and visualization to community decision-making processes for land use planning primarily at the watershed/landscape scale. He is responsible for teaching five courses including a studio which provides a service-learning experience for undergraduate students. The studio provides watershed-based land use planning assistance to Kentucky communities. He received his doctoral degree at The Pennsylvania State University, where his research focused on the form and function of community watershed organizations and their capacity building efforts in watershed management.

Richard H. Moore received his PhD in anthropology from the University of Texas at Austin in 1985. In 1985, he joined the Department of Anthropology at Ohio State University and in 1999 joined the Rural Sociology faculty, now part of the School of Environment and Natural Resources. He is director of the Environmental Science Graduate Program which has 90 faculty and 65 graduate students from across the university at OSU. He heads the Sugar Creek Research team of interdisciplinary scholars who have watershed grants from NSF, USDA, and USEPA. The team bridges the social and natural sciences and focuses on headwaters and participatory research.

Lois Wright Morton, PhD, is a Professor in the Department of Sociology in the College of Agriculture and Life Sciences at Iowa State University and interim director of The Leopold Center for Sustainable Agriculture. Dr. Morton's areas of research include civic structure, rural communities, natural resource management, and community-based watershed management. She has published extensively on civic structure and its relationship to agriculture, food systems, watershed management, and rural quality of life. Dr. Morton is project director of a number of state and federally funded water grants including USDA NIFA *Heartland Regional Water*

Coordination Initiative (2008–2012) and *Developing Local Leadership and Extension Capacity for Performance-driven Agricultural Environmental Management* (2008–2011), which have goals to transfer research into practice in watershed management and institutionalize a pilot educational program across the state of Iowa.

Jason Shaw Parker is a cultural anthropologist in the Department of Horticulture and Crop Science at The Ohio State University. His research focuses on multiple dimensions of sustainability and the environment including social aspects of water quality, agriculture, and food and food safety. Research foci include stakeholder group structure and land tenure and their effects on community-based watershed management. Other work examines links between food safety and sustainability with emphasis on farm size and marketing strategies. Dr. Parker has consulted with other watershed groups as an ecological anthropologist and taught courses such as Sustainable Agriculture and the Ecology of Food.

Max J. Pfeffer is International Professor of Development Sociology, Chair of the Department of Development Sociology, and Faculty Co-Director of the Community and Rural Development Institute at Cornell University. Max's research portfolio includes work on land use and environmental planning, community response to alternative energy development, and immigrant integration in rural communities. He recently published (with John Schelhas) *Saving Forests, Protecting People? Environmental Conservation in Central America.* Max's research program has been funded by the National Institutes of Health, the National Science Foundation, the U.S. Environmental Protection Agency, the U.S. Department of Agriculture's National Research Initiative and Fund for Rural America, and the Social Science Research Council.

Linda Stalker Prokopy is an Associate Professor of Natural Resources Planning at Purdue University. Her research, teaching, and extension focus on the social dimensions of watershed management. She is interested in how to motivate people to adopt water-friendly practices and how to get a diversity of people actively engaged in local natural resource decisions.

Theresa Selfa is an Assistant Professor in the Department of Environmental Studies at SUNY-ESF, Syracuse, NY. She has expertise in rural, environmental, and development sociology, with research experience in Brazil, the Philippines, Europe, and the United States. Her recent research has focused on consumer–producer networks in local food systems and on farmer environmental attitudes and land management practices in Devon, England, and in central Kansas. She is currently working on research examining the impacts of biofuels production on rural communities and environments in Kansas and Iowa, and on social sustainability certification systems for biofuels. Her work has been published in *Society and Natural Resources, Environment and Planning A, Journal of Rural Studies, Agriculture and Human Values,* and *Environmental Science and Policy.* She received her PhD in Development Sociology from Cornell University.

Richard Stedman is an Associate Professor in the Department of Natural Resources at Cornell University. His research and teaching focus on sense of place, environmental attitudes, private landowner behavior, community-based resource management, and the well-being of rural communities in the face of rapid social and environmental change. He received his PhD in Sociology from the University of Wisconsin – Madison in 2000.

Linda P. Wagenet is a Senior Extension Associate in the Department of Development Sociology at Cornell University. Her research area focuses on citizen involvement in environmental management, specifically at the watershed scale. Linda also has a research appointment in the Bronfenbrenner Life Course Center in the College of Human Ecology at Cornell and is a Faculty Fellow in the Cornell Center for a Sustainable Future. She holds an MS in Water Pollution/Water Chemistry from the University of California, Davis and a PhD in Adult Environmental Education from Cornell University.

Mark R. Weaver is a professor of political science at the College of Wooster and teaches courses in environmental policy, political theory, and constitutional law. In the past 10 years, he served 2 years as a Visiting Scientist with the Agroecosystems Management Program at the Ohio Agricultural Research and Development Center, Ohio State University, where much of his research focused on the Sugar Creek Watershed. His research interests focus on the formation and structure of grassroots watershed groups and the processes through which local stakeholders make decisions about resource use and conservation practices.

Jason L. Weigle is a PhD Candidate in Rural Sociology and Human Dimensions of Natural Resources and the Environment at The Pennsylvania State University. His dissertation research explores the emergence of social fields in response to development pressure from Marcellus Shale natural gas production in northern Pennsylvania. Other research interests include watershed management, land use planning, and community development.

Bill Welton is a professor of natural resources/environmental science at Haskell Indian Nations University. Prior to that, he worked as a resource manager in various positions within the USDA-Forest Service. Responsibilities included Forestry, Range Management, Wildlife Habitat, Fire Management, and Recreation on National Forests in Arizona, New Mexico, Arkansas and South Dakota. Born in Albuquerque, New Mexico, Welton graduated with a B.S. in Forest Land Management from Northern Arizona University in 1971 and obtained an M.S. in Forestry from University of Missouri-Columbia in 2000. His research activities have included switchgrass ecotype propagation and applied watershed management.

Jeff Zacharakis is an associate professor of adult education in the Department of Educational Leadership at Kansas State University. He is also a faculty associate of K-State's Institute for Civic Discourse and Democracy and advisory board member of K-State's Center for Engagement and Community Development. His research areas include community, organizational, and leadership development. Prior to K-State, he worked as a community development specialist for Iowa State University Extension and Northern Illinois University's Lindeman Center in both rural and urban settings and served on the administrative council for NCR-SARE.

Contents

Part I Pathways

1 **Pathways to Better Water Quality** 3
Lois Wright Morton and Susan S. Brown

2 **Citizen Involvement** 15
Lois Wright Morton

3 **Shared Leadership for Watershed Management** 29
Lois Wright Morton, Theresa Selfa, and Terrie A. Becerra

4 **Relationships, Connections, Influence, and Power** 41
Lois Wright Morton

5 **Turning Conflict into Citizen Participation and Power** 57
Jeff Zacharakis

6 **The Language of Conservation** 67
Jacqueline Comito and Matt Helmers

Part II The Data

7 **Measuring the Citizen Effect: What Does Good Citizen Involvement Look Like?** 83
Linda Stalker Prokopy and Kristin Floress

8 **Regional Water Quality Concern and Environmental Attitudes** 95
Zhihua Hu and Lois Wright Morton

 9 **Communities of Interest and the Negotiation
    of Watershed Management** 109
    Max J. Pfeffer and Linda P. Wagenet

10 **Upstream, Downstream: Forging a Rural–Urban
    Partnership for Shared Water Governance
    in Central Kansas** 121
    Theresa Selfa and Terrie A. Becerra

11 **Local Champions Speak Out: Pennsylvania's
    Community Watershed Organizations** 133
    Kathryn Brasier, Brian Lee, Richard Stedman,
    and Jason Weigle

12 **Community Watershed Planning: Vandalia, Missouri** 145
    Daniel Downing, Robert Broz, and Lois Wright Morton

13 **The Role of Force and Economic Sanctions
    in Protecting Watersheds** 155
    Kristen Corey and Lois Wright Morton

14 **Cross-Cultural Collaboration for Riparian Restoration
    on Tribal Lands in Kansas** 171
    Charles J. Barden, Lillian Fisher, William M. Welton,
    and Ryan Dyer

15 **Getting to Performance-Based Outcomes
    at the Watershed Level** 181
    Lois Wright Morton and Jean McGuire

16 **A Farmer Learning Circle: The Sugar Creek Partners, Ohio** 197
    Mark R. Weaver, Richard H. Moore, and Jason Shaw Parker

17 **Farmer Decision Makers: What Are They Thinking?** 213
    Lois Wright Morton

18 **Sustainability of Environmental Management – The Role
    of Technical Assistance as an Educational Program** 229
    Susan S. Brown and Chad Ingels

19 **Building Citizen Capacity** 247
    Susan S. Brown

Bibliography 257

Index 267

Contributors

Charles J. Barden
Horticulture, Forestry & Recreation Resource, Kansas State University,
2021 Throckmorton, Manhattan KS, 66506, USA
cbarden@ksu.edu

Terrie A. Becerra
Department of Environmental Sociology, 006 Classroom Building,
Oklahoma State University, Stillwater, OK 74078-4064

Kathryn Brasier
Assistant Professor of Rural Sociology, Department of Agricultural Economics &
Rural Sociology, Penn State University, 105B Armsby Bldg.
University Park, PA, 16802, USA
kbrasier@psu.edu

Susan S. Brown
Department of Sociology, Iowa State University,
303 East Hall, Ames IA, 50011-1070, USA
ssbrown@iastate.edu

Robert Broz
Extension Agriculture Engineering-Water Quality Program,
University of Missouri, 205 Agriculture Engineering Building,
Columbia MO, 65211, USA

Jacqueline Comito
Department of Sociology, Iowa State University,
103 East Hall, Ames IA, 50011-1070, USA
jcomito@iastate.edu

Kristen Corey
Iowa State University, 313 NW Bramble Road, Ankeny IA, 50023, USA
kcorey@dhs.state.ia.us

Daniel Downing
Agricultural Extension-Food Science & Nutrition,
University of Missouri, 226 Agricultural Engineering Building,
Columbia MO, 65211, USA
downingD@missouri.edu

Ryan Dyer
Prairie Band Potawatomi Tribal Council, 16281 Q Road,
Mayetta KS 66509, USA

Lillian Fisher
Haskell Indian Nations University, Lawrence KS 66046, USA

Kristin Floress
College of Natural Resources, 800 Reserve Street,
University of Wisconsin-Stevens Point, Stevens Point WI 54481, USA

Matt Helmers
Department of Agricultural and Biosystems Engineering, 219B Davidson Hall,
Iowa State University, Ames IA 50011-1070, USA

Zhihua Hu
Department of Sociology, Iowa State University,
403B East Hall, Ames IA, 50011-1070, USA
zhihuahu@iastate.edu

Chad Ingels
Iowa State University, Agronomy Extension, 201 E. Clark Ste 113,
Fayette IA 52142, USA

Jean McGuire
Department of Sociology, Iowa State University,
317B East Hall, Ames IA, 50011-1070, USA

Gerald A. Miller
Iowa State University, 2150 Beardshear Hall, Ames IA, 50011-2046, USA

Richard H. Moore
School of Environment and Natural Resources, 201 B Thorne,
OARDC-Wooster, The Ohio State University, Wooster OH 44691, USA

Lois Wright Morton
Department of Sociology, Iowa State University,
317C East Hall, Ames IA, 50011-1070, USA
lwmorton@iastate.edu

Jason Shaw Parker
Department of Horticulture and Crop Science, The Ohio State University,
233A Howlett, 2001 Fyfee Court, Columbus OH, 43210, USA

Max J. Pfeffer
Department of Development Sociology, Cornell University,
267 Roberts Hall, Ithaca NY, 14853, USA
mjp5@cornell.edu

Linda S. Prokopy
Department of Forestry and Natural Resources, Purdue University,
195 Marsteller Street, West Lafayette IN, 47907, USA
lprokopy@purdue.edu

Theresa Selfa
Department of Environmental Studies, SUNY-ESF 107 Marshall Hall,
Syracuse, NY 13210-2787, USA
tselfa@esf.edu

Linda P. Wagenet
Department of Development Sociology, Warren Hall Cornell University,
Ithaca NY, 14853, USA

Mark R. Weaver
College of Wooster, 130 Kauke Hall, 400 E University Street,
Wooster OH, 44691, USA
mweaver@wooster.edu

William M. Welton
Haskell Indian Nations University, Lawrence KS 66046, USA

Jeff Zacharakis
Kansas State University, 326 Bluemont, Manhattan KS, 66506, USA
jzachara@ksu.edu

Part I
Pathways

Chapter 1
Pathways to Better Water Quality

Lois Wright Morton and Susan S. Brown

It is accepted practice that solutions to water quality problems involve the physical and natural sciences as well as engineering and a variety of technologies. Often overlooked is the human factor – the social sciences of human perceptions and actions, social relations, and social organization. Humans, like all plants and animals, are a species in nature, affected by changes in water, air, soil, weather, and other natural elements. They are also creators of their environment with their continuous constructing, destructing, and reconstructing of the world about them. Busch et al.[1] call it "making nature." Humans have made our creeks, streams, rivers, and lakes what they are today. Some actions are deliberate decisions, like the Clean Water Act legislation or negotiated trade-offs for economic, social, or political gain. Many actions are not purposeful and lead to unintended and unexpected results. All actions intended or unintended have consequences that put the quality of our water resources at risk or place in motion protective measures. It is the human capacity to think and act that is the source of polluted and degraded waters. This same capacity also offers hope for finding new pathways for solving increasingly complex water problems.

This book describes and evaluates the human social actions occurring across the United States to solve the persistent and difficult problem of nonpoint source (NPS) pollution. The US Environmental Protection Agency (EPA) identifies pollution from urban and agricultural land transported by precipitation and runoff (NPS) as the leading source of water impairment in the United States.[2] Our focus is on how citizens affect water impairment individually and collectively as they interact with

---

[1] Busch, L., W. B. Lacy, J. Buirkhardt, D. Hemken, J. Moraga-Rojel, T. Koponen, and J. de Souza Silva. 1995. *Making Nature Shaping Culture*. Lincoln: University of Nebraska Press.

[2] *The Quality of Our Nation's Water*. http://www.epa.gov/305b/.

L.W. Morton (✉)
Department of Sociology, Iowa State University, 317C East Hall, Ames IA 50011-1070, USA
e-mail: lwmorton@iastate.edu

L.W. Morton and S.S. Brown (eds.), *Pathways for Getting to Better Water Quality: The Citizen Effect*, DOI 10.1007/978-1-4419-7282-8_1, © Springer Science+Business Media, LLC 2011

water daily. They have both rights to use and responsibility for conserving, protecting, and sustaining this valuable and vulnerable resource. We build on the social science evidence that links public deliberation to scientific knowledge about environmental problems.[3] We examine the citizen effect, the many ways people engage science, technology, and each other to identify and solve their watershed problems.

There is much to be learned from the social sciences as we search for ways to better protect our water resources. If significant water quality improvements are to occur, citizens must get involved.[4] Citizens are agricultural producers, rural non-farm residents with a quarter acre or hundreds of acres, and urban dwellers from small towns to big cities. They are the local girl-scout leader, the bank manager, the farm parts store clerk, the science teacher, the church deacon, the garbage truck driver, the auto technician, the health care assistant, the water plant operator, the mayor, the news reporter, and the farmer. They are the people who live, work, and play in a watershed. *The Citizen Effect* is about the processes and consequences of citizens' public engagement with water and each other and the actions they take to make their waters better.

Extent of the Problem

Nonpoint source pollutants delivered from across the landscape are the number one cause of impaired waters in the United States.[5] Although NPS pollution is diffuse, its ultimate source is readily understood as rooted in the day-to-day actions and management decisions of all citizens, urban and rural. Collectively, farmers are responsible for major NPS pollution impacts in many regions because agricultural practices require intensive management of much of the landscape. US Geological Survey scientists report that agricultural sources contribute more than 70% of the nitrogen (N) and phosphorus (P) delivered to the Gulf of Mexico via the Mississippi

---

[3] http://www.epa.gov/305b; Sabatier, Paul A., W. Focht, M. Lubell, Z. Trachtenberg, A. Vedlitz, and M. Matlock (eds). 2005. *Swimming Upstream: Collaborative Approaches to Watershed Management*. Cambridge: The MIT Press; Morton, Lois Wright and Steven Padgitt. 2005. "Selecting Socio-Economic Metrics for Watershed Management." *Environmental Monitoring & Assessment* 103:83–98; Morton, Lois Wright and Chih Yuan Weng. 2009. "Getting to Better Water Quality Outcomes: The Promise & Challenge of the Citizen Effect." *Agriculture and Human Values* 26(1):83–94.

[4] Fischer, F. 1993; 2005. *Citizens, Experts and the Environment: The Politics of Local Knowledge*. Durham: Duke University Press; Dietz, Thomas and Paul C. Sterns (eds). 2008. *Public Participation in Environmental Assessment and Decision Making*. Committee on the Human Dimensions of Global Change, Division of Behavioral and Social Sciences and Education. National Research Council of the National Academies. Washington: The National Academies Press http://www.nap.edu

[5] National Water Quality Inventory Report to Congress, http://www.epa.gov/305b

River Basin.[6] Further, corn and soybean cultivation contributes 52% of N, while P originates mostly from animal manure on pasture and rangeland (37%), followed by corn and soybeans (25%), and other crops (18%).

The most recent National Water Quality Inventory prepared by 50 states, the District of Columbia, 5 territories, 4 interstate commissions, and 5 Indian tribes reports that, in addition to nutrients such as N and P, other major causes of NPS impairment are siltation, bacteria, and oxygen-depleting substances. Nationally, US cropland covers about 25% of the total land area in the lower 48 states, or 400 million acres.[7] Data collected from 1992 to 2003 reveal 13% of US streams and 20% of groundwater wells have nitrate concentrations greater than the drinking water standard of 10 ppm; and 85% of streams have phosphorus concentrations in excess of the EPA federal criterion of 0.1 ppm and 13% have concentrations of 0.5 ppm or higher.[8]

Expert science, engineering, new technologies, and financial incentives have been the focus of public policy and the managerial tools of choice for responding to the problem of NPS pollution. US environmental conservation efforts in agriculture began with the soil conservation movement in the early twentieth century. A national soil erosion service and a system of giving grants to farmers to pay for soil management practices were established in the 1930s. When water quality emerged as a primary environmental concern in the period 1960–1970, it was shown that silt, nutrients, and chemicals delivered to water bodies by soil erosion were a principal cause of agricultural NPS pollution. As a result, water quality protection and improvement in agricultural landscapes has continued to be addressed based on the policies and programs originated for soil conservation. These strategies have been useful for understanding pollution causality and developing technical tools for assessment, prevention, restoration, and remediation actions for impaired lakes, streams, rivers, and wetlands. However, these strategies are missing the human social factors necessary to achieve sustainable success.

The physical sciences with a strong funded research base have built knowledge about ecological processes and offered technical interventions to impairments. Funding for the sociological sciences to build social knowledge and applications to watershed management has lagged behind. This shortcoming is being recognized as new guidelines for research proposals in natural resources research require researchers to integrate the human component with the physical sciences. Many of our authors in this book have been recipients of these grants and report out the first

---

[6]Alexander, Richard B., Richard A. Smith, Gregory E. Schwarz, Elizabeth W. Boyer, Jacqueline V. Nolan and John W. Brakebill. 2008. "Differences in Phosphorus and Nitrogen Delivery to the Gulf of Mexico from the Mississippi River Basin." *Environmental Science & Technology* 42(3):822–830.

[7]USDA Natural Resource Inventory; USDA Economic Research Service.

[8]Heinz Center. 2008. *The State of the Nation's Ecosystems 2008: Measuring the Lands, Waters, and Living Resources of the United States.* The H. John Heinz III Center for Science, Economy, and the Environment. Washington: Island Press.

wave of their findings in their chapters. The focus of their work is to increase our understanding of how citizens learn or do not learn about their natural environment and reasons they ignore issues or choose to act together to solve common problems.

Several emerging issues have increased the importance of citizen and community ownership of water problems. One is the continuing difficulty in promoting voluntary adoption of conservation practices, even with significant public cost share. Addressing producers' reluctance is a social issue as much as an economic issue. Another issue is the increased importance of pollutants, like nitrogen, which are not well controlled by sediment management. Sustainable improvement in management of these pollutants challenges public conservation efforts to integrate social science and education with existing programs. A third issue is the changing pattern of land ownership and its implications for environmental management decisions. The increasing regulation or proposed regulation of agriculture to reduce NPS pollutants will also have far-reaching impacts on individual producers and their communities.

Citizens as a Resource for Watershed Management

This book, *Pathways for Getting to Better Water Quality: The Citizen Effect*, proposes to connect sociological theories of citizen engagement and civic structure with empirical findings. Our intent is to offer science-based guidance for interventions that include the development of relationships among landowners, agricultural producers, community residents, and agency technical staff as they undertake planning and management in local watersheds.

Surface water is generally a publicly owned resource that can be used for ordinary private purposes. Groundwater ownership and use and surface water rights are becoming more complex as demand for water threatens to exceed available water resources. The management of US water resources has historically been government agency directed under an expert managerial model. A shift to increased public involvement in the early twentieth century resulted in legislation that required public notice prior to agency rule making. More recently, governments at many levels and citizens are recognizing that the multiple interests of the public are insufficiently represented through the managerial model. As a result, they are seeking ways to move beyond public notice and public hearings to active citizen participation in water resource management decisions. Beierle and Cayford[9] write "a fundamental challenge for administrative governance is reconciling the need for expertise in managing … with the transparency and participation demanded by a democratic system."

---

[9] Beierle, T. C. and J. Cayford. 2002. *Democracy in Practice: Public Participation in Environmental Decisions*, p. 3. Washington: Resources for the Future.

Citizens are an untapped resource in our efforts to solve water quality problems. Many environmental issues, including water quality, are "wicked problems"[4] – those with no easy solutions and wherein resolution is often temporary and imperfect. Wicked problems are not well suited to a managerial approach because they involve social values and political judgments. Resolution requires citizens to engage each other and negotiate solutions congruent with public and personal goals and norms. Resolution also requires a public will to invest personal and public resources and actions that sustain and improve the quality of water.

The theoretical underpinnings of the citizen effect are civic engagement and the building and strengthening of a local civic structure. Civic engagement is the public deliberations and actions that occur when people come together to respond to shared public concerns.[10] Civic structure consists of the formal and informal groups, organizations, institutions, and social relationships built by citizens that help the community deal with issues in the public commons.[11] Therefore, *The Citizen Effect* is about what can happen when ordinary people connect with other ordinary people to understand and solve the local problems of NPS water pollution. Citizens may not agree on how to solve a water-related problem. They may not even agree on how a problem is defined. However, if they can acknowledge that an issue is important and if they are willing to talk to each other about the issue, empirical evidence shows that there is the potential to find ways to deal with the issue.

Government agencies, private organizations, and community leaders have a major role in creating public spaces and finding ways to structure the public environment so citizen conversations and actions can occur. Many entities can contribute to education, defined as the change in individuals' ideas and actions that result from their involvement in participatory learning experiences. The intent of *The Citizen Effect* is to provide a scientific framework and empirical evidence to document the many ways people engage each other to make sense of and solve shared watershed concerns. This information in turn can guide technical watershed specialists, government agency professionals, Extension educators, and community leaders as they seek to better understand citizen involvement and find ways to engage citizens.

Place-Based Decision Making

The sociological examination of watershed management and citizen involvement is public sociology by its very nature. Our goal is to document and transfer the knowledge that sociology is building about citizen-stakeholders to watershed specialists

---

[10] Almond, G. and S. Verba. 1989. *The Civic Culture Revisited.* Newbury Park: Sage Publications; Putnam, R. D. 1993. *Making Democracy Work: Civic Traditions in Modern Italy.* Princeton: Princeton University Press.

[11] Morton, L. Wright. 2003. "Civic Structure." pp. 179–182 in *Encyclopedia of Community: From the Village to the Virtual World* edited by Karen Christensen and David Levinson, Thousand Oaks: Sage Publications.

and community leaders, i.e., to put our findings into the mainstream public discourse. The discourse itself changes how citizens view their water resource and their willingness to participate in watershed management rather than address a single acreage or farm in isolation. Discourse also changes how expert water specialists view the capacities of citizens to learn about their watershed and the roles they are willing and able to take in managing it.

Government-sponsored efforts to assess natural resource problems, generate solutions, and implement and evaluate local management actions are unavoidably fragmented because of political boundaries. The watershed however is a uniquely "natural" unit for citizen engagement in problem solving. A social structure of place framed around the natural resource base rather than political boundaries fosters the building of common vision, understandings, and goals and the development of effective intervention strategies. This premise is bolstered by Weber's[12] claim that "The emergence of hundreds of rural, place-based, grass-roots ecosystem management (GREM) efforts across the United States constitutes a new environmental movement that challenges the fundamental premises of existing natural resources and public lands institutions."

Citizen involvement in decision making around shared resources varies extensively, from apathy to passion and from politically appointed blue ribbon committees to established civic and sectoral associations to newly formed grassroots groups with a burning environmental cause. Experience has shown that the mere inclusion of citizens or citizen groups does not guarantee better outcomes; it may lead to inaction and stalemates or exacerbate poor economic and social conditions. The conditions under which citizens are asked to participate in watershed management decisions influence how water problems are addressed. Further, the knowledge, experiences, goals, beliefs, and values citizens bring to public discussions affect how science and technological solutions are understood and accepted. The practice of public sociology that shapes those conditions and recognizes those knowledge and belief systems has the potential to change not only the social relationships within a watershed but the water outcomes experienced locally and downstream.

Making Space for Citizens to Participate

The intended reader of this book includes scientific and technical experts, government agency natural resource professionals, extension agricultural educators, and community leaders who are championing water quality and acting as catalysts for changing land use practices. Watershed specialists bring expert technical skills to watershed management but may be unfamiliar with the social relationships and the decision-making processes farmers use in adopting/or not adopting conservation

---

[12]Weber, E. P. 2000. "A New Vanguard for the Environment: Grass-roots Ecosystem Management as a New Environmental Movement". *Society & Natural Resources* 13:237.

and land use recommendations. Involving citizens in watershed management is new and untried in many publicly funded projects designed to monitor, assess, plan, and reduce sediment and nutrient loads in officially designated impaired waterbodies.[13] Some agency professionals and technical experts are skeptical that ordinary citizens can grasp the science behind watershed assessment and adaptive management decisions. Alternatively, they feel they lack the training, time, and resources to educate the public in these areas or they do not view working with the public as central to the mission of their agency. A few are threatened when their expertise is unappreciated or not accepted by farmers who challenge their assumptions, or refuse to adopt, or are uninterested in adopting, proven conservation practices. Extension agricultural specialists acknowledge the value of engaging their stakeholders but often concentrate their efforts at the field and farm levels and miss opportunities to utilize the collective efforts of the community in addressing the big picture of local watershed impairment. Community leaders know they have knowledge and skills to contribute but become frustrated when their ideas and experiences are not sought out or accepted as valid by watershed experts. The research applications in this book are intended to bridge the knowledge gap between expert and citizen and to provide empirical support for citizen roles in water management.

Watershed planning and management involve complex technical, social, and political decisions. As we invest in scientific and technical resources, it is clear that the most intractable problems for watershed management remain those associated with the human dimension – targeting individuals, promoting adoption, and working with communities. This book will share some of what social scientists have learned about working with citizens and the influences they can have on water issues. Our target is agricultural NPS pollution and watershed management and planning at local and regional levels. However, the theories and the findings presented relating to citizen participation and public deliberation, the development of shared leadership, and citizen collective actions demonstrate how the power of science and technology, when understood and used by citizens, can lead to better management outcomes in any public watershed effort.

Outline of the Book

The book is divided into two sections. Section I: Pathways presents theoretical and philosophical foundations for why citizen involvement is valuable and how citizen influence affects local watershed management. Chapter 1 sets the context for citizen engagement and US water quality concerns. Chapter 2 proposes theoretical frameworks that link civil society, democracy, citizens, and social pressure to management of shared water resources. The Flora Agroecosystem Management Model

---

[13]EPA 303d list http://www.epa.gov/owow/tmdl/.

of social control, for example, is a structural explanation of the roles of government, markets, and civil society, and how they approach the problem of solving NPS pollution. The Catalytic Influence of Local Champions Model shows how exposing a few key people with a conservation frame of mind to science and new technologies can leverage civic structure to accomplish social and field/farm/watershed level outcomes. Case studies in subsequent chapters apply these models and other social theories to illustrate the pathways citizens use to get to better watershed-wide outcomes.

Chapters 3 and 4 discuss the social mechanisms of shared leadership, collective actions, influence, and power and the tensions between professional expertise and democratic engagement as they apply to watershed decision making. Zacharakis, in Chap. 5, offers useful insights into the types of conflicts encountered in watershed management and how they can be turned into positive energy to increase problem-solving capacities. Successful watershed management must deal with the many sources of conflict that arise from differing goals and expectations among watershed residents as well as between regulatory agencies and citizens. Zacharakis offers five reasons conflict can lead to positive watershed outcomes and cautions against suppressing public conversations when disagreements are present. He asserts that the key to successful watershed group development is creating a welcoming public environment where participants feel safe in expressing dissenting or supporting viewpoints.

Section I culminates with Comito and Helmers' call for re-languaging conservation to reflect social value shifts and to create a stronger culture of conservation. When conservation beliefs and land stewardship practices become cultural norms, citizens motivate each other to protect their shared natural resource base. Iowa Learning Farm listening sessions with farmers, state Department of Natural Resources (DNR), and USDA Natural Resources Conservation Service professionals reveal the impact of community, culture, and peer influence on perceptions of the environment and willingness to participate in building a culture of conservation.

Section II offers qualitative and quantitative data as evidence of citizen perceptions and understandings about water and how citizens have effectively engaged in public dialogues and actions. These chapters range from how to measure the citizen effect, applications of performance-based outcomes, and technical assistance as an educational program, to examples of citizen involvement in Kansas, Missouri, New York, Ohio, Pennsylvania, and Iowa watersheds. In Chap. 7, Prokopy and Floress ask what does good citizen involvement look like? They note that many hierarchies of participation have been developed during the last 40 years to help understand the depth of citizen involvement. These hierarchies help us think about what types of citizen participation can be most helpful in making watershed projects successful. The lowest level of participation is frequently co-optation; really a type of non-participation in which citizens are invited to participate only by being informed of the agenda already adopted by project organizers. In contrast, at the higher levels of involvement, citizens are active in project development and leadership. Successful watershed projects are those that can foster participation at the higher levels of these hierarchies. In addition to the depth of involvement, Prokopy and Floress assert that we also need to think about the breadth of participation. That is, we are not only concerned with how people participate, but who participates. Frequently participation in watershed management is limited to a small segment of the population that may not be representative of the larger population. In this chapter, the authors suggest ways in which both the depth and breadth of participation can be measured within a watershed project and offer evidence from watersheds in the Midwest to illustrate these indicators.

Perceptions of state-level water quality, environmental attitudes, and changes in environmental behaviors are the topic of Chap. 8. Hu and Morton analyze survey data from a stratified random sample of people's perceptions of their ground and surface water quality in 36 states. They find that views on the function and use of the environment as well as demographic characteristics are associated with changing behaviors such as frequency of watering lawns; using pesticides, fertilizers, or other chemicals on lawns; and how home yards are landscaped.

Chapters 9 and 10 examine the rural–urban interface as public drinking water supplies are impacted downstream by agricultural practices. Pfeffer and Wagenet tell the story of New York City (NYC) and Upstate New York farm communities as they negotiated the 1997 New York City Watershed Memorandum of Agreement. This agreement, signed by approximately 40 upstate towns and villages, environmental groups, the USEPA, New York State, and New York City serves as a blueprint for the city's watershed management strategy for west-of-Hudson sources. In 1990 NYC's proposed regulatory actions aroused strong opposition from upstate rural watershed residents who feared that economic development would be stifled, property values would drop, and the local tax base would be eroded. For the NYC Watershed Agreement to come about, organizations that were new and more inclusive had to form. These organizations had to encompass relations between upstate watershed communities and between the communities and NYC. This chapter identifies key organization factors that made the historic NYC Watershed Agreement possible.

In Chap. 10, Selfa and Becerra present a case study of Cheney Lake watershed in central Kansas. They chronicle two complementary storylines – one is the urban Wichita community's need for clean drinking water and their willingness to pay farmers to change their land use and conservation practices to ensure a safe water supply. The second is the story of an agricultural community recognizing their larger role as caretakers of the soil and water resources on which their livelihood relies. This case study illustrates how rural and urban citizens can identify mutual benefits and set common goals, create a plan, and act together. Their joint efforts create shared governance structures that build and sustain watershed management capacities.

Brasier et al. (Chap. 11) define community-based watershed organizations (CWOs) as liaisons between individual users/managers of land and water resources and the communities and governmental agencies that have authority over those resources. They find that CWOs have the potential to provide crucial links among local and state organizations, private and public organizations, resource developers and users, and individual landowners. In their chapter, they distill interviews with leaders from 28 CWOs in Pennsylvania to describe strategies they use to achieve goals related to water quality as well as building skills and abilities of both individuals and communities. CWOs become a local, place-based gathering point for mobilizing resources in the form of community organizing, partnerships, funding, and technical resources related to watershed improvement. Initially, CWOs focus on water quality improvement. Ultimately they find that to improve water quality, CWOs need to increase the capacity for action at a local level by creating collaboration opportunities, increasing stakeholder leadership skills, and adapting and transforming ecological as well as social and political knowledge.

Community watershed planning is the focus of Chap. 12. Downing et al. discuss EPA's nine-element framework for watershed planning and the regulatory expectations when a watershed is classified as impaired and is required to develop a total maximum daily load (TMDL) plan. Their work shows that involving citizens in TMDL development is time consuming but necessary if land use practices are to change and be sustained. When Vandalia, Missouri public drinking water supplies registered Atrazine levels that exceeded health standards, the community mobilized to solve the problem. This chapter takes the reader through the EPA nine-element process and illustrates citizen roles in each step.

While the dominant theme of most chapters in this book is voluntary social pressure to solve impaired water issues, in Chap. 13, Corey and Morton document the role of force and economic sanctions in achieving water goals. The Saline Wetlands of Lincoln, Nebraska are presented as a case study showing roles of planning and zoning staff and an application of the Endangered Species Act to a wetland ecology. A fragmented sense of community and suburban sprawl into farmland provides the context for regulatory and enforcement agency actions, citizen involvement, and capacity to protect wetlands.

Chapter 14 by Barden et al. uses 10 years of collaboration on water quality-related projects between the Prairie Band of the Potawatomi Nation (PBPN) and the two land grant universities in Kansas, Haskell Indian Nations University and Kansas State University, to illustrate cross-cultural collaboration for riparian restoration on Tribal lands in Kansas. This case study uses the framework of Tribal sovereignty and self-determination to show the impact of technical water specialists' hands-on work with PBPN members to restore riparian woodlands and build local capacity to improve surface water quality on their reservation in northeast Kansas. Traditional Indian belief prioritizes water as the most sacred of elements. This provides the basis for building an ongoing collaboration including streambank stabilization, riparian restoration, and watershed bio-assessment activities.

Performance-based environmental management applied to agricultural production and collective actions is the theme of Chap. 15. Morton and McGuire use the Hewitt Creek Model and interviews with farmers to tell the story of an on-farm performance-based environmental management initiative in a Northeast Iowa watershed. Farmers share leadership in monitoring their impaired streams, set performance goals for their watersheds, undertake activities on their own fields and farms, and motivate farmer neighbors to adopt conservation practices.

As a result of this collective effort, farmers have strengthened their own personal knowledge of their watershed and increased their management skills to co-produce both profitable agriculture and environmental improvement. Field- and farm-level management tools, such as a phosphorus index (P Index), stalk nitrate tests, and the Soil Conditioning Index (SCI) are used by this farmer group to create a holistic picture of what is happening in their watershed and to develop watershed-level management responses.

The value of farmers meeting and talking together about water issues has not only been proven in Iowa, but also Ohio. In Chap. 16, Weaver et al. describe a farmer learning circle and the impacts it has had in one of Ohio's most impaired watersheds. The Sugar Creek Partners project illustrates what a successful farmer-led watershed group can accomplish. Farmer-led groups are sites of information exchange, demonstration trials, and learning experiences as well as sources of social support, collective action, and community leadership. The Upper Sugar Creek farmer group has used neighbor-to-neighbor influence to create an 8-mi. buffer strip to reduce pollution from agricultural runoff and protect against stream bank erosion, and to promote watershed conservation management practices as part of good stewardship of the land.

In Chap. 17, Morton asks what are farmer decision makers thinking and then offers a brief glimpse of farmers' mental maps and general frames of reference that guide their management practices. She discusses prescriptive and facilitative transfer of scientific knowledge and the effect on adoption of a system of managing for both agricultural productivity and agroecosystem services. Farmer interviews, surveys, and focus groups are used to illustrate the farmer decision-making perspective and to show that it can be distinct in specific ways from the expert professional perspective. Morton suggests that understanding the farmer viewpoint provides a starting point for educators to better support farmer goals and expectations while targeting water quality outcomes at farm and watershed levels. The chapter ends by challenging educators and natural resource professionals to help farmers reconstruct their values and beliefs in ways that support conservation management as a system of farming.

Technical assistance as an educational program can be a source of sustaining farmer environmental management practices claim Ingels and Brown in Chap. 18. They illustrate how technical specialists' responsiveness, listening, and willingness to learn from the farmer can build trust and increase farmer requests for assistance with environmental best management practices. The planning process for best management practices in turn offers many opportunities to provide education and decision tools to the farmer as the person who is expected to implement those plans and expand upon their environmental benefits. In this chapter, Ingles, an extension watershed technical specialist, discusses how combining the expert role with educational support brings farmers and landowners into a management partnership that leads to more sustainable environmental outcomes.

We conclude the book with a discussion on building community capacities to better integrate and balance scientific and expert knowledge, local citizen experiences and knowledge, and diverse social values and goals. The US EPA and the

United States Department of Agriculture, as well as numerous federal and private funding organizations, are increasingly requiring community involvement in watershed planning and implementation proposals. The intent is to get citizens substantively engaged in watershed management decisions and build community capacity. The challenge for the public sector is to facilitate citizen awareness of water resource issues and offer opportunities for learning that will influence community norms and motivate local actions.

Chapter 2
Citizen Involvement

Lois Wright Morton

Citizens and Civil Society

Civil society and its citizens are the centerpiece of this book. The concept of civil society in democracy is centuries old and is distinct from the roles of government and market sectors.[1] Human society is held together by shared normative under-standings and guided by public discourse and moral persuasion rather than external coercion.[2] The conditions of democracy confer equal status on members of society and provide the normative acceptance of idea and opinion exchange and collective deliberations in the public arena.[3] Thus, citizens in a democracy are persons who associate with other persons, creating communities of cooperation that engage in conscious actions for the public good.[4] When a person assumes the role of citizen, he or she places the public good over personal self-interest and accepts both mutual rights and obligations.

Our authors explore the roles of citizens and how they influence the management of local watersheds and in turn affect water quality outcomes. All watershed com-munities are place bound, that is, they consist of groups of people bound together

---

[1] Knox, T. M. 1967. *Translation of Hegel's Philosophy of Right*. Oxford [1821]: Oxford University Press; McNicoll, Geoffrey. 1995. *Demography in the Unmaking of Civil Society*. The Population Council. Working Paper No. 79. Canberra: Australian National University; Janoski, Thomas. 1998. *Citizenship and Civil Society*. United Kingdom: Cambridge University Press.

[2] William M. Sullivan. 1999. "Making Civil Society work: Democracy as a Problem of Civic Cooperation." Chapter 2 in *Civil Society, Democracy, and Civic Renewal* edited by Robert K. Fullinwider. New York: Rowman and Littlefield Publishers.

[3] Cohen, Jean L. 1999. "American Civil Society Talk." Chapter 3 in *Civil Society, Democracy, and Civic Renewal* edited by Robert K. Fullinwider. New York: Rowman and Littlefield Publishers.

[4] Barber, Benjamin R. 1996. *Jihad vs. McWorld*. New York: Ballantine Books.

L.W. Morton (✉)
Department of Sociology, Iowa State University, 317C East Hall, Ames IA 50011-1070, USA
e-mail: lwmorton@iastate.edu

L.W. Morton and S.S. Brown (eds.), *Pathways for Getting to Better Water Quality: The Citizen Effect*, DOI 10.1007/978-1-4419-7282-8_2,
© Springer Science+Business Media, LLC 2011

by the land that drains water into the physical flow of common streams, rivers, lakes, and other bodies of water. People who live and work in a watershed are by default citizens of that watershed. However, although all expect to benefit from water, few actually consider their responsibilities to protect and manage it as a shared resource. One reason for forming community watershed groups is to leverage the latent citizen role and affirm the expectation that citizens have obligations to their watershed. The groups and organizations of civil society provide the capacity for leadership to develop and if "harnessed and nurtured, can transform local democracy and reshape" communities.[5]

First, I portray water as the public commons and discuss the challenges society has in selecting the "right" outcomes they wish to pursue collectively. Then sociological explanations for why citizen involvement in watershed management can transform beliefs and management practices in the watershed are discussed. Lastly, an intervention model is offered for obtaining social and field/farm practice outcomes that can lead to improved watershed outcomes.

Water as the Public Commons

Our natural resource base is a multiple-use, shared resource, and as such presents issues of how to manage the public commons.[6] Rivers, lakes, streams, and wetlands are critical sources of life, energy, and economic and social well being. The conservation and protection of these limited resources involves public and private land use decisions, agricultural practices and policies, natural resource rights, public supplies and disposal, lifestyle and consumption behaviors, and allocation of moral and financial responsibilities. Although a shared resource, there is limited agreement on the magnitude of water degradation, the extent of resource loss, irreversibility, and how personal and/or communal resources should be invested.

The global frame of reference that scientists, state and federal government agencies, politicians, and environmental activists bring to public discussions often vary a great deal on how they perceive water resource issues and what public policies should be enacted. These global views contrast with local citizen frames of reference based in their personal experiences, identities of place, and how their lives are affected by their unique local water bodies. Viewpoints among local citizens also

---

[5] Ball, Colin and Barry Knight. 1999. "Why We Must Listen to Citizens." In Chapter 2 in *Civil Society at the Millennium*, edited by Kumi Naidoo, 20. West Hartford: Kumarian Press.

[6] Ravnborg, Helle M. and M. del Pilar Guerrero. 1999. "Collective Action in Watershed Management-experiences from the Andean Hillsides." *Agriculture and Human Values* 16: 257–266; Steins, Nathalie A. and Victoria M. Edwards. 1999. "Synthesis: Platforms for Collective Action in Multiple-use Common-pool Resources." *Agriculture and Human Values* 16:309–315; Morton, L. W. 2003c. "Civic Watershed Communities." Chapter 8 in *Walking Towards Justice: Democratization in Rural Life* edited by Michael M. Bell and Fred T. Hendricks with Azril Bacal, 121–134. Research in Rural Sociology and Development Vol. 9, Amsterdam: JAI/Elsevier.

vary widely with local topology, beliefs and understandings of nature, and place in
the life cycle influencing their perceptions and actions. Because watersheds are
systems with effects beyond local, goals and strategies for managing this resource
are negotiated among local and regional, national, and sometimes international
interests.[7]

The call for greater citizen involvement in watershed management has increased
the complexity and changed the structure of water quality problem solving.[8] Public
involvement spans multiple geographic scales and viewpoints about water use and
functions and the range of possible actions. Although increased opportunities exist
for citizens to become involved in water issues, the outcomes of citizen participation in management are not always predictable and have not been systematically
documented. Case examples abound that demonstrate that citizen engagement does
not always lead to mission-directed agency preferred or even expected outcomes.
The reasons for these disjunctures are multiple – human nature and social actions
are complex and social scientists are just beginning to propose and test theories to
identify patterns that can be used to guide interventions.

In some instances, the inclusion of ordinary citizens in what historically has
been an expert-driven environmental management model has led to a vision to
restore and protect the resource base and institute protective land use policies and
actions.[9] At the same time, citizen involvement can also increase conflict and con-
testation of rights, responsibilities, and solutions.[10] Citizens are not a monolithic
block but rather have a wide range of political and environmental positions in favor
of, neutral to, and against resource uses and protective actions. The public negotiation of these positions to create sufficient agreement to allocate resources and make
decisions can become divisive and contentious.

Citizen versus citizen counteractions have leveraged change, increased the threshold of expectations for protection, and introduced new perspectives for negotiating
water outcomes. Sometimes citizen involvement and passions become bitter, leading

---

[7] Parisi, D., M. Taquino, S. M. Grice, and D. A. Gill. 2004. "Civic Responsibility and the
Environment: Linking Local Conditions to Community Environmental Activeness." *Society &
Natural Resources* 17:97–112.

[8] Dietz, T. and Paul C. Stern (ed). 2008. *Public Participation in Environmental Assessment and
Decision Making*. Committee on the Human Dimensions of Global Change, Division of
Behavioral and Social Sciences and Education. National Research Council of the National
Academies. Washington: The National Academies press http://www.nap.edu.

[9] Weber, Edward P. 2000. "A New Vanguard for the Environment: Grass-roots Ecosystem
Management as a New Environmental Movement." *Society and Natural Resources* 13:237–259.

Morton, L. W. and S. Padgitt. 2005. "Selecting Socio-Economic Metrics for Watershed
Management." *Environmental Monitoring and Assessment* 103:83–98.

[10] Zacharakis, Jeff. 2006. "Conflict as a Form of Capital in Controversial Community Development
Projects." *Journal of Extension* (October) 44(5):5FEA2. (http://www.joe.org/joe/2006october/
a2.shtml).

to polarized viewpoints and stalemates that prevent problem-solving strategies from being discussed and implemented. These conflicts often pit economic needs against quality of life expectations of different sectors. In contrast to places with active public dialogues, there are communities where citizens are apathetic to water issues. This apathy can stem from unawareness of the critical nature of the issues, a lack of leadership to mobilize and inspire, a fear or distrust of outside agencies (e.g., US Environmental Protection Agency (EPA) or Department of Natural Resources (DNR)) imposing solutions, and/or an unwillingness to engage in public issues.

The challenge for elected leaders, natural resource educators, and agency professionals mandated to protect water quality is to build broad local support for watershed management. Without citizen engagement and their voluntary support, it is difficult to resolve many water concerns. A first step in building citizen support is agreement on public goals and socially preferred outcomes.

Getting to "Right" Outcomes

US watershed communities are subject to federal laws for water quality set forth by the 1972 Clean Water Act and subsequent revisions. This is the "stick" behind local and state actions, the force of social control that compels and enforces the protection and restoration of US waters via the public agency, the EPA. Farm Bill legislation implemented by the United States Department of Agriculture (USDA) during the past several decades has provided the "carrots," individual farm economic incentives to practice conservation measures that can lead to better water quality. These rules and regulations and economic incentives have provided some motivation for changing how society behaves, but they are not enough. Flora[11] asserts that water management "is more than the technical operation of water systems"; it includes the governance of the community and how the community thinks about its water issues.

A growing body of literature supports the idea that watershed management is more than a public agency responsibility and should include the people most affected by water decisions – the citizens of a watershed.[12] In recent years, government agencies,

---

[11] Flora, Cornelia Butler. 2004. "Social Aspects of Small Water Systems." *Journal of Contemporary Water Research and Education* 128:6.

[12] Weber, Edward P. 2003. *Bringing Society Back In: Grassroots Ecosystem Management, Accountability, and Sustainable Communities*. Cambridge: MIT Press.

Fischer, Frank. 2005. *Citizens, Experts, and the Environment: The Politics of Local Knowledge*. Durham: Duke University Press.

Koontz, Tomas M., Toddi A. Steelman, JoAnn Carmin, Katrina Smith Korfmacher, Cassandra Moseley, Craig W. Thomas. 2004. *Collaborative Environmental Management: What Roles for Government?* Washington D.C.: Resources for the Future.

Beierle, Thomas C. and Jerry Cayford. 2002. *Democracy in Practice: Public Participation in Environmental Decisions.* Washington D.C.: Resources for the Future;

Sabatier, Paul A., Will Focht, Mark Lubell, Zev Trachtenberg, Arnold Vedlitz, and Marty Matlock. 2005. *Swimming Upstream: Collaborative Approaches to Watershed Management.* Cambridge: MIT Press.

voluntary organizations, and citizens themselves have called for more citizen involvement, but it is not always clear what they are expecting to happen. Social science theories of civil society and the role of citizens in water resource management can offer guidance in selecting interventions that include public participation. Two questions direct much of the current research in this area: *How does connecting people to new information and to each other in their watersheds change the political and social landscape sufficiently to influence the physical conditions of our waters? What are the mechanisms and under what conditions is citizen involvement most effective?*

Several sociological theories explain how the citizen process works and why involving citizens can be a productive strategy in watershed management. However, before applying our theories and findings to an intervention plan, we must first define the dependent variable in our proposed questions. We must answer the value-laden questions: What outcomes are we expecting and what are the "right" outcomes to evaluate the effectiveness of citizen involvement? Once we know our preferred outcomes, we are in a position to ask how we can best get there. Without a clear definition of preferred outcomes, it is difficult to develop a roadmap of where citizens fit and what we expect of them.

"Right" Outcomes are Socially Defined

Herein lies a knotty problem ... society must define and agree on which outcomes are the "right" ones before choosing specific actions. Society is many different people with many different viewpoints. What viewpoint is the "right" one? How can we build some kind of agreement so we respond in a timely manner to increasing threats to water quality? Some biophysical scientists and strong pro-environmentalists assert that we must restore our waters to their original pristine conditions, without human pollutants. Others will say, we cannot afford to restore it to that level and besides we don't drink from all water bodies, so a non-body contact standard is good enough for some waters. What outcomes are "good enough?"

The choice of "right" outcomes must integrate scientific and local knowledge, public attitudes, beliefs, and normative expectations about water and land use functions. In discussions that center on nonpoint source pollution, for example, agricultural land use is the context for many public conversations about water conditions and goals for water quality. In rural landscapes, land with good soils has historically had a primary function of producing profitable agricultural products for food, feed, fiber, and fuel. Our growing knowledge of ecosystem processes and problems of water scarcity and poor water quality has led to the addition of a second function, that of producing ecosystem services. Ecosystem services include all of the resources and processes of the natural world vital to human health and livelihoods, such as clean drinking water, nutrient cycling, water purification, reducing the magnitude of flooding, carbon storage, and safe water-based recreation. These factors often are taken for granted as free public goods because they lack a formal market. Now that we are beginning to recognize that many of these services are threatened and limited, "right" outcomes must include social decisions about tradeoffs among them, and scientific knowledge can provide only part of the answer.

The role of choosing the right inputs, outputs, and outcomes for water has in recent history been left to scientists, technology professionals, mission-directed agencies, and politicians. Sometimes their decisions are grounded in science and at other times they are guided by political necessity. For example, in 2004, the Iowa Department of Natural Resources proposed weaker standards for sewage treatment plants for cities. "Regulators admitted that the limit hasn't been enforced in its 20-year history"[13] and recommended a standard they could enforce. At issue was the strictness/weakness of the chloride and other pollutant standards and the amount of salt released into local streams.[14] Environmental groups wanted rules that were more stringent; city sewage-treatment workers, elected officials, and representatives of dairies and other industries said the cost of meeting the limits was too high. Public standards and enforcement are frequently negotiated among sectors that differ greatly in their expectations and their available resources to address water concerns.

Getting to the "right" outcomes for public resources such as water requires building a shared knowledge base about the physical and social conditions of the watershed and negotiation of priorities among local citizens and the sectors they represent. Civic discovery processes rebalance the "bias of bureaucratic expertise toward narrow, technocratic solutions" with "equally valid claims of those who must live with the outcomes."[12] Scientists who think developing incredibly compelling data is all that is necessary are stymied to find local stakeholders may have different perceptions and values that discount science.[12] Further, because watersheds are systems with effects beyond local, outcomes and strategies for getting there must also be negotiated among many levels of geographical, political, and social interests. This suggests that the social structure and scale at which interactions take place are important factors in achieving desired outcomes.

Thus, the answer to the question, what are the "right" outcomes, is that it depends. It depends on *who* is engaged in the processes of managing the water body, *what* their special interests are, and *how* they negotiate the sectoral values and priorities placed on water issues. It depends on *what kind* of public conversations about their water body citizens who live and work in the watershed have with each other. It depends on how well public agencies and private organizations with production and environmental missions are able to convey the known science and help citizens learn and evaluate what they've learned in light of their own personal beliefs about the environment and the social norms of the society in which they live. In other words, what is "right" is socially constructed, and, hence, there is a need for sociological analysis of water quality issues.

---

[13]Beeman, Perry. 2004. "Panel Votes to Let Limited on River Pollution Stand." *The Des Moines Register*, p. 3B, March 16.

[14]Excessive salts are toxic to some aquatic plants, insects, and other organisms.

Mission-directed public agencies often have different goals and expectations than private commodity and environmental organizations. Further, individual citizens and groups to which they belong are likely to have differing goals and preferred outcomes for their own lands and those of others. This situation is "why" the citizen effect is so crucial to solving water quality concerns. Water is a public, shared resource. Managing it requires collective and public efforts to minimize the sectoral differences, to build a common vision, and to negotiate solutions and create a plan that has some level of unity in its implementation.

How does Citizen Involvement Get Us to "Right" Outcomes?

Social Theory

Groups in civil society exist to supply their members with some desired common good.[15] The challenge is how to foster and encourage the formation of watershed communities of cooperation that consciously act for the public good – the protection of water resources. This common good can only be attained if members comply with the rules necessary for the production of that good. There are several ways to obtain compliance: coercion, compensation, and obligation.

The Flora[11] Agroecosystem Management Model (Fig. 2.1) of social control offers a macro-structural framework for how government, markets, and civil society create rules and obtain compliance in managing the ecosystem as a public good. Federal and state regulators typically have depended upon two strategies, force and economic sanctions to meet agency and publicly legislated goals to protect water and the environment. Force includes the power to set regulations, and to police and enforce them through economic fines or shutting down the polluting system. The EPA has been assigned this power through the 1972 US Clean Water Act. In addition, state and local agencies and governments have enacted a variety of home rule and land use laws that are used in conjunction with federal laws. Many of these rules designate appropriate standards of conduct and practice. Underlying the enactment of these rules is the acceptance by citizens that they were created on their behalf for the protection of the public good. The violation of socially set rules brings a sanction.[15]

Local zoning, wetland protection regulations, and flood plain designations are intended to match land use to topology, erodible soils, and areas prone to seasonal flooding. Guidelines are often unevenly applied and are dependent upon agency priorities to finance policing and upon a political willingness to enforce. When enforcement does occur, economic sanctions in the form of fines on individuals, cities, or businesses are used as punishment for current violations and to deter future violations. Flooding and drought issues involve both water quantity and flow

[15] Hechter, Michael. 1987. *Principles of Group Solidarity*. Berkeley: University of California Press.

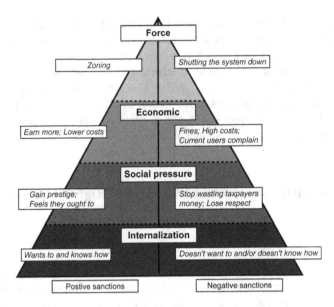

Fig. 2.1 Flora's agroecosystem management model

standards needed to protect life and property, and to guard against negative economic impacts on livelihoods. State Emergency Preparedness Offices require flood plain plans, and state and local zoning rules often institute safeguards against flooding. Force and economic sanctions are usually based in top-down judgments about what is best for the local watershed.

Economic incentives such as cost share and rent land payments are positive sanctions that encourage private resource investments and a willingness to change individual or corporate practices. However, these incentives often must continue indefinitely for change to be sustained. When 10- or 15-year Conservation Reserve Program (CRP) acreage enrollment expires, farmers often return the land to row crop production unless additional economic incentives are offered. Further, depending on commodity market prices, farmers may lose money when land is enrolled in lower-paying conservation programs. Economic incentives require public willingness to tax themselves so landowners can be paid to use their lands in ways that protect the watershed.

Dependence on force and economic sanctions alone to achieve water quality and adequate water quantities are limited by motivation and sustainability. These sanctions are costly to monitor and enforce and often do not motivate people to voluntarily engage in practices that lead to better water outcomes. Further, they do not create or reaffirm a conservation ethic nor reproduce values or reasons for maintaining high water quality. Once force and economic sanctions are removed, individuals and corporations have little to no incentive to undertake protective measures unless they feel social pressure or have developed internal values to protect water resources.

Social Pressure and Internalization

While force of regulation and economic incentives are necessary elements in meeting water goals, they are insufficient to not only achieve goals but also to sustain long-term actions. Citizen obligation to act in the public interest rather than personal interest is derived from internal and external pressures. The base of the Flora pyramid uses social pressure and internal beliefs and values to obtain compliance. In the formation of watershed groups, the citizen obligation to act on behalf of the watershed is leveraged, thereby stimulating the latent citizenship that has the potential to change the landscape.

For some farmers, internal beliefs and systems of beliefs underlie the adoption/lack of adoption of conservation practices.[16] In other cases, the social network, the example of others, and the social cultural expectations exert pressure.[15] Beliefs are people's perceptions about the world and how it works. They arise from a combination of experiences, scientific facts, observation, myths and stories, learned behaviors, and unverified assumptions that help people to make sense of the world. Beliefs become knowledge when verified. Objective (factual) and subjective (built from personal experience) knowledge intermingle and are not sharply differentiated but are influential in how issues are framed.[17] Exposure to new relationships, ideas, and technologies can lead to reconstruction of current mental maps about how the world works and to new identities as conservationist farmers.[18] Knowledge influences personal actions as it becomes internalized; often changing how problems are framed and causing reevaluation of taken-for-granted assumptions.[17]

Values are deeply embedded ideas and orientations about what is important, and right and wrong. Attitudes are an assessment or subjective evaluation based on facts and experience filtered through beliefs and values. Taken together, knowledge, beliefs, values, and attitudes are sources of motivation to retain the status quo or change behaviors.

Social Pressure

"Groups influence their members by subjecting them to a variety of obligations to act in the corporate interest and by ensuring that these obligations will be fulfilled."[15] At the local watershed level, particularly those at Hydrological Unit Code (HUC)

---

[16]McCown, R. L. 2005. "New Thinking About Farmer Decision Makers." Chapter 2 in *The Farmer's Decision: Balancing Economic Successful Agriculture Production with Environmental Quality*, edited by Jerry L. Hatfield. Ankeny: Soil and Water Conservation Society.

[17]Innes, Judith E. 1994. *Knowledge and Public Policy: The Search for Meaningful Indicators*, 2nd ed. London: Transaction Publisher.

[18]Coughenour, C. Milton. 2003. "Innovating Conservation Agriculture: The Case of No-till Cropping." *Rural Sociology* 68(2):278–304.

12^{19} and smaller, people know each other and already have built some networks and connections that have the potential to be directed toward water concerns. Repeated interactions within a watershed community can help generate local norms of reciprocity between members, because each learns gradually to expect that the other will cooperate[3] to respond to a shared concern about their water. The formation of watershed groups builds on preexisting relationships and strengthens them by explicitly identifying common goals and utilizing public discussions and social pressure to transfer new ideas and knowledge as well as conservation beliefs and attitudes about the watershed.

The theory of civic structure[20] offers a partial explanation for the effectiveness of social pressure in the reproduction of conservation practices. The civic structure of the watershed is the network of multiple and diverse relationships among individuals and organizations and how they participate in shared decision making about public resources.[21] Attributes of this structure are the communication flows,[22] stakeholder groups, and the dynamic and changing social pressures derived from generalized norms and laws that reinforce community benefit over self-interest actions. Civic structure influences the capacity and ability of local leaders in formal and informal organizations to exert leadership on behalf of the community interest and to guide group decision making in the allocation of resources and joint actions to solve common problems. The characteristics and strength of a local civic structure influences the amount of social pressure placed on individuals and groups to conform, challenge, or ignore public concerns about the water commons.

Embedded in civic structure definitions are normative expectations that people and organizations will act in ways that benefit the community because they are citizens who share a common place. In the United States, there are legal laws rooted in the practice of democracy that protect public participation and the right to challenge how government and markets manage the resources of society. Although there are legal protections for tolerating dissent and competing ideas, local communities vary in their tolerance for competitive viewpoints and challenges to those that dominate local decision making. A strong civic structure is characterized by extensive and substantive communication flows among residents. Information exchange and public dialogue are necessary for framing public issues and arriving at politically and socially acceptable goals (preferred outcomes) and acceptable solutions perceived to accomplish common goals.

[19] HUC12 watersheds are 10,000-40,000 acres or 15-62 mi2.

[20] Morton, L. W. 2003a. "Small Town Services and Facilities: The Influence of Social Capital and Civic Structure on Perceptions of Quality." *City & Community* 2(2):99–118.

[21] Morton, L. W. 2003b. "Civic Structure." In *Encyclopedia of Community: From the Village to the Virtual World,* edited by Karen Christensen and David Levinson, 179–182. Thousand Oaks: Sage.

[22] Yankelovich, Daniel. 2001. *The Magic of Dialogue: Transforming Conflict into Cooperation.* New York: Touchstone.

Democracy and Civic Involvement

In a democratic country such as the United States, citizens have responsibilities to identify and engage in solving public problems. The values of democracy and the experience and practice of democratic social pressure are central to explaining why citizen involvement can be a successful strategy[23] for addressing environmental issues such as water.

A number of social scientists link characteristics of the term "civic" to democracy in a causal multidirectional manner.[24] Theories of democratic stability are grounded in the role of civic culture.[24] For Almond and Verba,[24] civic culture includes consensus on the legitimacy of political institutions, high tolerance levels for a plurality of interests, and mutual trust among citizens. Warren[25] traces the development of democracy along with civic virtue in associations whose goals are public and inclusive social goods created by cooperation. Civic virtue, the consistent pursuit of the public good at the expense of individual and private ends, leads to better economic outcomes, says Putnam.[26] Skocpol and Fiorina[27] note that the social trust and civic engagement of Putnam is based on a causal chain: individuals interact with others in face-to-face settings (build social capital), they learn to work together to solve problems, and social trust is created, which in turn leads to good public policies, better economic development, and efficient public administration.

An Intervention

What we know about social pressure and internal beliefs and values suggests an intervention model that uses two to three key watershed citizens as catalysts to get to social and farm level outcomes that can lead to improved watershed conditions. Figure 2.2, the Catalytic Influence of Local Champions, illustrates the dynamic impact a few key people can have on reconstructing values and beliefs and leveraging the civic structure to influence other farmers and land managers and transform their watershed. The basis for the model is providing opportunities for the beliefs, experiences, and cultural knowledge of key people to be influenced by scientific knowledge.

[23]Boyte, Harry C. 2008. *The Citizen Solution: How You Can Make a Difference.* St. Paul: Minnesota Historical Society.

[24]Almond, G. A. and S. Verba. 1989. *The Civic Culture Revisited.* London: Sage Publications; Putnam, R. D. 2000. *Bowling Alone: The Collapse and Revival of American Community.* New York: Simon & Schuster.

[25]Warren, M. E. 2001. *Democracy and Association.* Princeton: Princeton University Press.

[26]Putnam, R. D. 1993. *Making Democracy Work.* Princeton: Princeton University Press.

[27]Skocpol, T. and M. P. Fiorina. 1999. *Civic Engagement in American Democracy.* Washington: Brookings Institution Press.

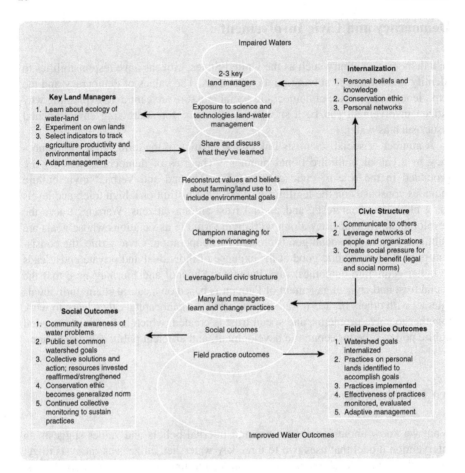

Fig. 2.2 The catalytic influence of local champions

Exposure to science and technologies are used to aid farmers in learning about land–water ecological relationships and become a motivation for experimentation on their own lands. Science and technology interventions that facilitate farmer learning help them make sense of their own experiences and knowledge and lead to adaptations in their management practices.[16]

The initial starting point of the intervention is the identification and recruitment of two or three key land managers who have control or influence over agricultural land management decisions. In addition, these key land managers must have personal internal belief systems that are open to viewing themselves as not only producing agricultural products but also some of the environmental benefits described as ecosystem services. Once this core group is convened, they become referents for each other, sharing what they have learned as they are exposed to new ideas, science, and technologies, and evaluating this information against their own knowledge. It is the willingness to learn and the commitment and passion of these key people as they build new personal knowledge that has the potential to transform[28] the watershed.

Public discussions about land and water conditions and the collective exchange of information aid in the reconstruction of values and beliefs[18] that are necessary for a shift in land management for both profitable production and ecosystem services. While these key landowners are not likely to use the term ecosystem services, exposure to science and technologies provides support for increasing understandings of land–water–animal–plant relationships. Peer-to-peer exchange of information and personal experimentation offer opportunities to evaluate what they are learning, and reinforces the systems approach and the need for their active engagement and continuous adaptation of management practices for long-term production and environmental protection.

The collective effort among several key land managers has a spillover effect as they naturally champion what they have learned and the new management practices they have put in place to ensure higher water quality.[29] Bandura[30] finds that most behaviors that people engage in are learned, either deliberately or inadvertently, through the influence of example. The formation of a citizen watershed group provides a public forum for knowledge exchange and learning from the example of others while creating a horizontal network of trust and cooperation. Relationships among those of equal status and power are what generate good social capital[3] and in turn a strong civic structure. Social pressure on other land managers to adopt new practices takes several forms under these conditions: persuasion, reputation, imitation, boasting rights, and peer normative sanctions. As the norm of managing for better water quality on personal fields becomes generalized beyond the farm level, some non-conforming land managers will feel socially pressured to shift practices even if they do not actively participate in the watershed group.

Social Outcomes

To understand the role of social pressure in challenging and reconstructing values and beliefs that mobilize the watershed community, it is useful to identify the social outcomes that can change the physical landscape. The box on the left of Fig. 2.2 summarizes a number of social outcomes that are the result of interactions among personal internal beliefs and knowledge and civic structure.[29] The social outcomes of community awareness, setting common goals, and collective solutions are also the outcomes of the democratic process. These in turn can lead to the development

---

[28] Polanyi, M. 1958. *Personal Knowledge: Towards a Post-critical Philosophy.* Chicago: University of Chicago Press.

[29] Morton, L. W. 2008. "The Role of Civic Structure in Achieving Performance Based Watershed Management." *Society and Natural Resources* 21(9):751–766.

[30] Bandura, Albert. 1971. *Social Learning Theory.* Morristown: General Learning Press.

of a new community norm or reaffirmation of existing community norms that value conservation and the co-production of ecosystem services and agricultural profitability. A shared conservation ethic provides the foundation for community agreement in how public and private resources are invested. These resources are necessary to get to improved water outcomes. Sustaining this achievement requires community vigilance and continual collective efforts to adapt to social, economic, political, and ecosystem changes.

A portion of the variation among watershed communities in their capacities to address water issues can be attributed to how citizens get involved. In Sect. II, empirical evidence is offered in support of the catalytic role citizens play in changing community norms and ensuring that there are opportunities to practice public dialogue and information exchange. Not all communities will be successful in dealing with their water problems. Some will lack strong leaders with environmental priorities to catalyze and guide change. Many will need outside resources and agency technical support in addition to good leadership. Some communities need the threat of regulation and enforcement to move out of their comfort zone to discover that they can go beyond their current conditions. Some places have leaders who discourage and block public engagement and challenges to market-driven decisions on development in environmentally sensitive areas. Other places are simply laissez faire, allowing government rule makers and technical professionals to propose and implement solutions as they think is best.

The combination of all four elements of Flora's social control pyramid offers the potential to identify the right goals and to put in place practices that can support agricultural productivity and protect our vulnerable water resources. Neither force nor economic incentives have been completely effective in solving nonpoint source pollution. Further, if government were to take policing and enforcement and farm payments as the primary means of solving the problem, the taxpayer cost would be prohibitively high and result in minimum standards that assume static conditions in a dynamic system. Social control strategies must also include citizen involvement and collective actions. Social pressure, the building of civic structure, and the understanding of social organizations and social–psychological formation of internal beliefs and values are the foundations for getting to better water quality outcomes.

A common vision and the creation of a unified plan for water resources begin with the geographical unit, the watershed and the citizens who live there. Framing the social structure of place around the natural resource base rather than political boundaries reduces the fragmentation that occurs throughout the processes of assessment, generation of solutions, and decision making, and the implementation and evaluation stages. Grass-roots ecosystem management (GREM), as Weber[9] documents, is grounded in the geography of the local watershed. Citizen involvement is at the heart of this environmental movement. The reasons locally led environmental management works are found in the sciences of sociology and anthropology. As we learn about social networks and connections, power and influence relationships, and the role of culture, we will have a deeper understanding of how citizen involvement works and what we can expect when people engage each other in solving the problem of the water commons.

Chapter 3
Shared Leadership for Watershed Management

Lois Wright Morton, Theresa Selfa, and Terrie A. Becerra

Grass Roots Leadership

Grass roots ecosystem management builds local partnerships with citizen leadership,[1] creates a common vision among stakeholders, and sets community goals that support appropriate planning, implementation, and monitoring approaches. Leach and Pelkey[2] find that effective leadership and management are key factors in successful watershed partnerships. Wheatley has described leadership not as a role but as a behavior.[3] Shared leadership occurs when multiple members of a watershed community engage in leadership behaviors simultaneously. Leaders are most effective when they are considered reliable, credible, and respected by the community,[4] with reputations that are transmitted through personal relationship networks. When community members become partners in solving water issues, these connections create webs of influence rather than chains of command. This approach mobilizes resources that include people traditionally left out of decision-making processes and generates collective problem solving.[3]

---

[1] Weber, E. P. 2000. "A New Vanguard for the Environment: Grass-roots Ecosystem Management as New Environmental Movement." *Society & Natural Resources* 13:237–259.

[2] Leach, W. D. and N. W. Pelkey. 2001. "Making Watershed Partnerships Work: A Review of the Empirical Literature." *Journal of Water Resources Planning and Management* (November/December) 378–385.

[3] Wheatley, M. 1999. *Leadership and the New Science.* San Francisco, CA: Berrett-Koehler Publishers; Morton, L. Wright. 2008. "The Role of Civic Structure in Achieving Performance Based Watershed Management." *Society and Natural Resources* 21(9):751–766.

[4] Goode, E. and N. Ben-Yehuda. 1994. *Moral Panics: The Social Construction of Deviance.* Oxford: Blackwell.

L.W. Morton (✉)
Department of Sociology, Iowa State University, 317C East Hall, Ames IA 50011-1070, USA
e-mail: lwmorton@iastate.edu

L.W. Morton and S.S. Brown (eds.), *Pathways for Getting to Better Water Quality: The Citizen Effect*, DOI 10.1007/978-1-4419-7282-8_3,
© Springer Science+Business Media, LLC 2011

Effective expert interventions to improve watershed management engage citizens in leadership development. The goal of leadership development is to stimulate people to think critically and innovatively as they seek new solutions to achieve current and future goals.[5] Capable leaders are critical in articulating and sustaining a vision for the watershed ecosystem and encouraging community self-organizing behaviors and practices. They provide key group functions such as creating trust, making sense of information, compiling and generating knowledge, linking stakeholders, and mobilizing broad support for change.[6] Also, they are sources of agenda setting and innovation. The development of citizen leadership includes building and strengthening a variety of skills: visioning, building trust, teamwork, group process and facilitation, decision making, and communication.

Leaders who have a vision for their watershed become "moral crusaders"[4] and champions for solving water problems. It is accepted science that small changes in an ecosystem can produce big effects.[7] This same principle can be applied to watershed management and the catalytic function leadership plays in transforming impaired waters.

Beyond Management: Leadership with Vision

Many public resources have been invested in research and development of environmental "better management" strategies that change land use practices in watersheds. However, promoting management change alone is doomed to fail unless land managers and owners and the community at large have leaders with a vision of what must be accomplished and why.[8] This effect is particularly clear in the case of nonpoint source pollutants that can only be controlled by changing the day-to-day management decisions made by many individual farmers. Leaders carry the vision that guides daily thinking and actions. Their vision gives purpose and meaning to watershed activities, compelling others to put aside self-interests and championing the restoration of the whole watershed as the goal. One Iowa farmer puts it this way,

[5] Allen, B. Lundy and L.W. Morton. 2006. "Shared Leadership Practices among Non-profits in Iowa." *Journal of Extension* December 44(6):1–12. (http://www.joe.org).

[6] Folke, C., T. Hahn, P. Olsson, and J. Norberg. 2005. "Adaptive Governance of Social-Ecological Systems." *Annual Review Environment Resources* 30:441–473.

[7] Strange, C. J. 2007. "Facing the Brink without Crossing It." *Bioscience* December 57(11):920–926.

[8] Dietz, T. and Paul C. Stern (eds). 2008. *Public Participation in Environmental Assessment and Decision Making*. The National Research Council of the National Academies The National Academies Press; Washington. (http://www.nap.edu).

> My vision would be every farmer in [my] … watershed doing a [conservation] practice, maybe at least one practice on every farm. That's my kind of thing, if every farmer would just try something to improve.[9]

Another Iowa farmer in the same watershed group adds his vision,

> Good water, you could almost drink out of it … just to have it cleaned up with a lot of fish in it, the kind a guy could eat.[10]

The vision gives the watershed community a sense of joint mission and purpose.[11] Leaders with vision provide the "energy behind every effort and the force that pushes through all the problems."[11]

The challenge for natural resource professionals and community members is to find those with vision who are willing to lead and to empower them to be effective in helping others to change landscape practices in support of better water quality outcomes. In this chapter, we share the experiences of watershed specialists, agricultural and natural resource educators, and community leaders as they developed their vision; built trust and strengthened relationships; and moved from a few key leaders to watershed wide change. We use examples from case studies in Kansas and Iowa to illustrate leadership concepts. Iowa quotes are drawn from audiotaped and transcribed interviews held in 2005 with 12 farmers and local Extension educators who are part of a farmer-led watershed group. Kansas data are based on the transcription of a focus group held in June 2008 with 11 Kansas Watershed Restoration and Protection Strategy (WRAPS) watershed specialists. WRAPS specialists provide technical support and community development guidance to farmers and community watershed groups.

Finding Those Who Can Lead

Watershed leadership must be developed. While some individuals have natural leadership characteristics, these skills must be nurtured and encouraged or they will not be available to make a difference in the watershed. Leadership is about influence.[11] All watersheds have multiple networks of social relationships and influence. These webs of influence will vary by the number of years lived in the watershed, number of kin, topography and terrain, occupation, type of agriculture,

---

[9]Farmer #7 in interviews with NE Iowa Farmers 2005 from a Farmer Group that has met for two years using a performance driven environmental management model facilitated by Cooperative Extension.

[10]Farmer #6 interviews with NE Iowa Farmers, 2005.

[11]Maxwell, John C. 1993. *Developing the Leader within You.* Nashville: Thomas Nelson Publishers.

watershed size, and the size of owner parcels within the watershed.[12] Some social relationships will be strong, others weak, a few hostile, and many simply neighbors-at-a-distance. Each person has influence on others and is in turn influenced. One Iowa farmer comments, "Dave ... had a lot of influence on the parts that I use in my planter...."[13] Another Iowa farmer talks about his neighbor and what he learned from watching how he managed his farm,

> ...a very knowledgeable private individual ... worked in-depth with ISU for studies [comparing] no-till versus till. And after watching him for several years ... [I got] involved in it myself.[13]

A Kansas watershed specialist leverages these spheres of influence to get farmers to adopt conservation practices. He says, "What I find works best with producers is having another producer. If you can find one, you can get two or three others to change."[14]

Leaders emerge from the group because their views make sense given what the group and individuals need.[3] One Kansas watershed specialist says,

> ...there are certain individuals within a community that are recognized by other citizens in that community as being ... leaders.[15]

An Iowa farmer describes a person he considers a leader in his watershed,

> Well, I look at [another farmer] ... he's pretty level-headed, doesn't go off at the other end.... He's pretty smart on farming, so he's kind of the mentor that we look up to.... He's the one that I confide in.... He's probably the biggest influence right now.... He wants to see it [the watershed project] work and he wants to see the best for farmers too. So that's what means a lot.[16]

Many different kinds of people can provide leadership depending on the context and situation.[17] Often, watersheds don't have one leader but multiple ones who step up to lead when their relationships with others and the situation push or pull them to the forefront. Kansas WRAPS watershed specialists elaborate,

---

[12]Prokopy, L., K. Floress, D. Klotthor-Weinkauf, and A. Baumgart-Gertz. 2008. "Determinants of Agricultural Best Management Practice Adoption: Evidence from the Literature." *Journal of Soil and Water Conservation* 63(5):300–311.

[13]Farmer #2 NE Iowa Farmer Interviews 2005.

[14]Transcript line 201–205. Kansas WRAPS focus group 2008.

[15]Transcript line 340–343. Kansas Watershed Restoration and Protection Strategy (WRAPS) specialists provide technical support and education to farmers and community watershed groups. Quotes from these specialists are drawn from a focus group held June 6, 2008 at Kansas State University, Manhattan, KS. Eleven specialists (7 men and 4 women) participated in the 2-h discussion on their strategies for involving citizens in local watershed decision making and management.

[16]Farmer #4 NE Iowa Farmer Interviews 2005.

[17]Janov, J. 1994. *The Inventive Organization: Hope and Daring at Work.* San Francisco: Jossey-Bass.

> ...you have a core group of people who you've gained their trust and who are coming to the meetings. And they understand that the assessment is science; it's not regulation. It's not regulators going out and knocking on doors. It's a scientific assessment and these are the conditions, these are our problems.[18]

This is:

> ...a committee of local people ... that are willing to be voices ... to go out to work with the other people. That's what we've got ... that's been immensely successful ... all farmers that live in the watershed. Since 1994, they've met every month – once a month – at a meeting and you know they enjoy talking water. They said their best meetings are at the side of the road – stop and talk to a farmer in a pickup truck and that's the way you know we've gotten the watershed on board ... that nucleus group that was willing to go out and work [with other farmers]...[19]

Maxwell[11] cites three conditions needed for leadership to emerge. The person must have seen leadership modeled, have learned to lead by training and practice, and have self-discipline and a desire to lead. For leadership to emerge in the watershed, citizens need opportunities to observe good leaders, a chance to practice leading, and to view their leading as important to achieving watershed goals. Leaders use their social relationships and networks to influence others. However, achieving better water quality outcomes requires more than someone willing to lead. Maxwell[11] is quite blunt about this influence, "leadership is the ability to obtain followers." Watershed leaders can take their followers in many different directions, in support of conservation practices or against them. To solve the problem of water quality, leaders must believe there is a problem, understand the sources of degradation, and be able to articulate a vision that can lead to change. It is a leader's passion for water and a willingness to commit energy, time, and resources that engages others to want to let the vision guide their land use management decisions.

Developing a Vision

Those who study leadership suggest one of three reasons drives change, "people change when they *hurt* enough they *have* to change; *learn* enough they *want* to change, *receive* enough they are *able* to change."[11] These reasons correspond with Flora's model[20] of agroecosystem management that identifies force, economic pressures, social pressures, and internal values and beliefs as guiding natural resource decision

---

[18]Transcript line 561-566 Kansas WRAPS focus group 2008.

[19]Transcript line 495-507 Kansas WRAPS focus group 2008.

[20]Flora, C. Butler. 2004. "Social Aspects of Small Water Systems." *Journal of Contemporary Water Research and Education* 128:2.

making.[21] The first reason, "hurt enough they have to change" is top-down, when big stick management–regulatory forces pressure farmers to change management practices or suffer extreme consequences. Citizens may also be "hurt enough" if serious water degradation threatens their livelihood, drinking water supply, or safety.

The last reason "receive enough they are able to change" is the financial incentive strategy used by conservation programs of the federal Farm Bill legislation and other state programs. Agencies that manage these incentives claim that the profit motive is what drives farmer behavior. They believe that people fail to do the "right thing" for the environment because they perceive they can't afford to adopt different practices without public funds to share the cost. To others, these incentives appear as bribes. In any case they often fail to recruit individuals to change practices. Both regulation and incentives, the so-called "sticks" and "carrots," share the characteristic that they are external pressures designed to control individual landowner behaviors and practices toward expert-set land use and water goals.

In contrast, the reason "learn enough" depends on internal pressures to change. This motivation means new information, experiences, and insights cause citizens to reconstruct their personal knowledge and belief base. When personal beliefs and attitudes shift, citizens reexamine the implications of past decisions and actions and voluntarily readjust management practices because of what is learned. Leaders with a vision create conditions for people to learn and self-organize in response to new information.

At issue for many farmers is *who* decided what the stream or lake use should be and *who* decided a local water body is impaired? And whose vision is it what a local watershed should look like and who is deciding what activities should be undertaken to achieve the vision? Most of these decisions are external to those who live in the watershed and often are made with little or no consent or input from local residents. The assessment measures are not understood, the remediation plan does not feel local, and the prescribed strategies for water improvement are someone else's, not the persons whose land management practices are being targeted. Government agency specialists and educators may have their own goals and activities for change, but they often have no followers. This is not their fault, they generally have no training or experience in leadership or cultivating local leaders. However, the result is farmers who have not caught the vision and are unconvinced of their need to change.

A Vision for Impaired Waters

The dilemma for engineers and technical professionals, agricultural educators, and community development specialists is to move water resource protection forward. The US Clean Water Act requires each state to designate specific uses for its lakes, streams, and rivers – such as swimming, fishing, drinking, or maintaining healthy

---

[21]Morton, L. Wright and Chih Y. Weng. 2009. "Getting to Better Water Quality Outcomes: The Promise and Challenge of the Citizen Effect." *Agriculture and Human Values* 26(1):83–94.

populations of fish and other aquatic life – and then assess their quality based on federal or state standards for that use. When the water quality does not meet the water standard, it is considered impaired. Every state has a list of impaired waters, commonly called the 303(d) list named after section 303(d) of the Clean Water Act. Once on the list, the state must develop a water quality improvement plan for each impaired stream stretch, river, and lake according to the legislation and current regulations [http://www.iowadnr.com/water/watershed/impaired.html]. The proliferation of these impaired water bodies across the United States has resulted in scientific experts and government agencies setting the vision for change and the strategies for watershed restoration.

People will not follow agency personnel beyond their stated authority unless they gain legitimacy and permission to lead.[11] When compliance does occur, it is often a minimum level of effort rather than a vigorous response.[22] "Force, used in leadership, signifies a deficiency that can only be temporarily compensated.... If power resides in the follower, then effective leaders must first learn what matters to their followers."[22] They must have trust and permission to lead and to guide community leaders.

Trust and Quality Relationships

When public funds are provided to watershed management projects, such as the Kansas WRAPS program, they generally include a professional agency staff position or watershed specialist charged with some or all the steps of planning and implementation. The level of trust and quality of the relationship between influential leaders and the local watershed specialist affects the clarity of vision and ability of the group to undertake solutions. A Kansas WRAPS watershed specialist observes,

> We've had well over 90% of the farmers that we talk to, agree to and implement the Atrazine best management practices. So I think that in itself is a success there. But in addition to that, we've also been able to measure actual water quality improvements from doing those practices ... and I think the keys were some of the things that we've already talked about for that project ... the trust [we've built]. We ... were fortunate enough to be able to hire [someone who] lives in the watershed, farms in the watershed.... And so he knows the farmers, they know him, they trust him, they know ... that he knows what he's talking about, and when ... he goes to their farm and talks to them ... he's already past the trust thing ... and he's been extremely successful.[23]

[22]Daniels, Aubrey C. and James E. Daniels. 2007. *Measure of a Leader*, p. 12. New York: McGraw-Hill.

[23]Transcript line 949-960 Kansas WRAPS focus group 2008.

The importance of trust, local acceptance of the watershed problem, and a vision for how the community could respond cannot be overestimated. A Kansas WRAPS watershed specialist reports,

> I know our watershed had U.S. Geological Survey do a lot of the early on water quality testing for us. There was a certain suspicion of them because they had government tags on all their vehicles. It was interesting. There was a real turning point in our project ... there was a high school class where a teacher realized it was good to have his kids – a biology and chemistry class – out in the watershed. They went out; they did water sampling and water testing and when those kids got the same results that USGS had gotten, all of the sudden USGS is okay. Because the neighbors' kids said the same [that] USGS is saying.[24]

Leaders have followers when they are trusted. A good leader listens and attempts to understand the values, aspirations, and frustrations of those who live and work in the watershed. An experienced Kansas watershed specialist says,

> ...you have to listen to people first and you have to meet them where they're at ... you're talking about water or water quality and you really have to have the bigger picture of where they're coming from because nobody's functioning in a vacuum. And they may be perfectly willing to listen about something they need to change on their farm, but ... also – you have to be aware of, you know, they are dealing with crops and prices and economics and family.[25]

It is not unusual for a local group of farmers to convene a watershed group with the sole purpose of getting off the Environmental Protection Agency's (EPA) 303(d) impaired water list or to avoid regulation.[26] While in the short-run, the negative goal to avoid more regulation can drive producers to change management practices, it sets a low standard for change and reinforces anger and resentment toward external pressures to change. An "us" versus "them" division is created despite the common benefit that would accrue if management changes were undertaken and sustained. When scientists and agency professionals set the goals and vision for a watershed, citizens feel uninvolved and not responsible for the activities needed to achieve the vision. They will focus on making a living off their land, targeting agricultural production efficiency, and keeping environmental regulatory rules at bay.

A number of farmer watershed groups have successfully gotten beyond the goal of getting off "the list" and created a positive vision for how managing differently could help their bottom financials and their local water quality. This transformation occurs when local leaders are trusted, articulate a vision for their watershed, and influence others to adopt that vision and experiment with altering daily land use practices. Leaders keep before their neighbors and friends the vision of what their watershed should and could look like and the compelling reasons why it matters. Keeping the issue in the forefront takes persistence and commitment, which is one

---

[24] Transcript line 584-589 Kansas WRAPS focus group 2008.
[25] Transcript line 243-251 Kansas WRAPS focus group 2008.
[26] NE Iowa watershed interviews 2005.

reason local leaders must take up this task. It is easy for the community to lose sight of the goal as rural and urban citizens sort their way through competing priorities and the complexity of watershed management. One Kansas WRAPS watershed specialist[27] who works on developing community watershed groups reinforces this point, "When there's not ... a rallying point so to speak ... that can leave a void in your watershed project."[28] That void is caused by the lack of vision and reason for why citizens should invest time and energy on water issues.

Moving from a Few Key Leaders to Watershed-Wide Change

Collective citizen actions seldom just happen. Someone, often more than one person, has to facilitate the development of local leadership, support organizational development, and nurture[29] the learning process necessary for leadership to emerge and farmers to engage in watershed management. This is an important role for an outside specialist. Once a core group of people share a watershed vision, they are ready to take leadership for developing strategies to achieve that vision. Maxwell[11] calls this the production phrase of leadership. People throughout the watershed come together to accomplish a commonly held purpose and are ready to undertake concrete, results-oriented actions. When people engage each other to learn about their collective identity within the watershed, they begin to see how their personal patterns and behaviors contribute to the whole.[3] And, surprisingly, they are ready then to take responsibility for changing themselves.[3]

Finding the Trigger Points

It is one thing to recognize the need for engaging private citizens and accept that their personal interest and passion are needed for effective watershed decision making;[30] but the path to make it happen is not always easy. Kansas WRAPS watershed specialists recommend that you have to find the trigger point to get local attention and engagement. One specialist says,

---

[27] Kansas Watershed Restoration and Protection Strategy (WRAPS) specialists provide technical support and education to farmers and community watershed groups.

[28] Transcript line 474-485 Kansas WRAPS focus group 2008.

[29] Fischer, F. 2000. *Citizens, Experts, and the Environment: The Politics of Local Knowledge.* Durham: Duke University Press.

[30] Bonnell, J. E. and T. M. Koontz. 2007. "Stumbling Forward: The Organizational Challenges of Building and Sustaining Collaborative Watershed Management." *Society and Natural Resources* 20:153–167.

I think a lot of times for people to really get intimately involved with a project there has to be some kind of a trigger that really gets them interested ... and that's different for every watershed and every area....[31]

Wheatley[3] uses the analogy of the spider to illustrate how a threat or crisis to a living system can trigger efforts at self-repair. When a system is in trouble, the solution is to connect it to more of itself, in the spider's case, weaving more web. Similarly, when people realize their watershed is in trouble, they are often ready to work at both learning about the ecological system and strengthening their relationships with others to build a stronger system capable of solving its own problems. "The solutions the system needs are usually already present in it. If a system is suffering, this indicates that it lacks sufficient access to itself" writes Wheatley.[3] The system could lack information, might have lost clarity of identity, and/or might have troubled relationships. Wheatley continues, for change to occur, "the system needs to learn more about itself from itself."[3] This perspective suggests that outside experts cannot "fix" the system with a top-down effort. People must learn about their environment so they can recognize what is working and what is not working and take the needed steps to repair the system.

A variety of events can trigger citizens' attention. A Kansas WRAPS water specialist recalls,

[T]he thing that really got those people focused and motivated and organized was when a developer came in and wanted to put a reservoir in. You know, take thousands of acres of land and put in this reservoir. Well, that was the trigger ... got those people organized and motivated and into action.[32]

Another watershed specialist says each watershed community looks at issues in a different way,

...because sometimes the triggers are positive in a sense that it doesn't have to be an algae bloom or disaster. It's ... they're going to give you a $22 rain gauge and I'm going to do this because I love it you know.[33]

And one specialist chimes in, "Yeah taste and odor ... people started having taste and odor problems in their drinking water. That's the big time trigger in Kansas."[34] Once citizens learn the relationship between public drinking water taste and odor problems and algae blooms triggered by excess nutrients in water, they are ready to seek solutions.

---

[31] Transcript line 371–373 Kansas WRAPS focus group 2008.
[32] Transcript line 381–389 Kansas WRAPS focus group 2008.
[33] Transcript line 448–451 Kansas WRAPS focus group 2008.
[34] Transcript line 405–406 Kansas WRAPS focus group 2008.

Co-Managing and Leadership in the Watershed

Scientists and technical experts in conjunction with influential leaders can change the pathway of watershed restoration in three ways. The first is to help people connect to the basic identity of the watershed. This process is akin to setting a vision and begins with knowing their watershed address,[35] engaging in public dialog about water and land conditions, and learning what they would like those conditions to be. The second is to connect people to new information, emerging science, and technical alternatives. Information is dynamic and constantly changing. "[F]or a system to remain alive ... information must be continually generated."[3] Lastly, people must develop relationships with each other and their watershed. This approach means asking who is not involved in making the watershed work better, and how can we get them involved. Change in the watershed needs watershed experts willing to seek and empower citizen-leaders. Once citizens have made the vision for change their own, they will be ready to engage the scientist and watershed professional as partners in co-managing their water resource.

---

[35] Watershed address means the closest stream or water body to where a person lives and into which larger bodies it runs.

Chapter 4
Relationships, Connections, Influence, and Power

Lois Wright Morton

Influence and power are derived from the web of connections among people and the quality of these relationships.[1] As this idea suggests, and the wide experience of environmental advocates confirms, it is not a critical mass of people that solves problems but the critical connections among them that matter most.[1] In this chapter, relationships, connections, power, and influence are discussed as enabling processes used to engage citizens in learning about water, defining problems, and acting on those issues viewed as most critical. These processes promote participation among people, organizations, and communities in the setting and achievement of global and specific water goals.

"Problems of implementation are really issues of how to influence behavior, change the course of events, overcome resistance, and get people to do things they would not otherwise do."[2] In other words, implementing change is about influence and power. Both forms of persuasion are necessary to mobilize social support and resources. Power produces influence and influence produces power in a recurring cycle.[3]

People are not just political subjects, but also political actors whose choices and influence on others can reaffirm current conditions as well as become sources of change. Convening citizens to discuss and deliberate on water issues is only a starting point in getting to better land management and water quality outcomes. People must also be willing to act on what they learn and be so convinced that it matters that they use their power and influence to change others' beliefs, opinions, and behaviors.

---

[1] Wheatley, M.J. 1999. *Leadership and the New Science.* San Francisco, CA: Berrett-Koehler Publisher.

[2] Pfeffer, J. 1994. *Managing with Power: Politics and Influence in Organizations.* Boston, MA: Harvard Business Press.

[3] Willer, D., M.J. Lovagila, and B. Markovsky. 1997. "Power and Influence: A Theoretical Bridge." *Social Forces.* December 76(2):571–603.

L.W. Morton (✉)
Department of Sociology, Iowa State University, 317C East Hall, Ames IA 50011-1070, USA
e-mail: lwmorton@iastate.edu

L.W. Morton and S.S. Brown (eds.), *Pathways for Getting to Better Water Quality: The Citizen Effect,* DOI 10.1007/978-1-4419-7282-8_4,
© Springer Science+Business Media, LLC 2011

Survey and qualitative data collected in Iowa, Kansas, Missouri, and Nebraska are used to illustrate how relationships and connections among people and institutions influence land use management decisions. First, water as a political problem is discussed, followed by definitions and types of power and influence. The next section provides insights into how networks and connections give capacity to local watershed management, including individual influence, changes of mind, willingness to act, and observations on power and influence. Lastly, an informal political accounting system that can be used by community leaders and technical professionals is offered for analyzing power and influence relationships in local watershed management.

A Political Problem

Water pollution is a political problem. "A political problem is one in which you must get some other people to act or stop acting in a certain way to achieve a goal important to you."[4] The protection and restoration of lakes and reservoirs, rivers, streams, wetlands, and estuaries is a social goal many have embraced. This goal fosters the expectation that these water bodies should meet minimum water quality standards for their designated uses, including swimming, drinking, and the protection and propagation of aquatic life. Further, it is important to recognize that the standards and the designated uses are also political because their "official" definitions are creations of public agencies.

The achievement of water goals takes more than research in ecological sciences, and more than the development of water management technologies and the creation of mission-directed public agencies. It requires society to have the *political will* to act on the science of what is known, to apply those technologies shown to be effective, and to actively engage each other in changing personal behaviors and human-created institutions that are sources of water pollution. Resource-management decisions reflect human values and behaviors rather than physical biological conditions.[5] These decisions connect the natural environment to humans as both species in the environment and creators of it.

Personal networks are sources of power and influence and each person who lives in a watershed has some level of influence and power. Those who wish to guide watershed projects need to make sense of these networks and the key individuals within them. How important the water concern is to them and their position on the issues will determine whether they ignore, support, or block change. When two

---

[4]Coplin, W.D. and M.K O'Leary. 1972. *Everyman's Prince: A Guide to Understanding Your Political Problems*, p. 4. North Scituate, MA: Duxbury.

[5]Grumbine, R.E. 1997. "Reflections of 'What is Ecosystem Management?" *Conservation Biology* 11(1):41–47.

sectors place high priority on their natural resource base but have strongly divergent views, it is easy to polarize the community. As a result, people take sides rather than look for common ground for solving shared concerns. Understanding the power and influence structure of local watersheds and larger basins can provide a basis for bridging disagreements and strengthening those relationships that can have positive impacts on addressing water issues.

Power and Influence

The concept of power has many associated meanings, negative and positive, with impacts at different levels – individual, group, institutional, local, regional, national, and global. Power and influence issues have been a source of scientific inquiry for centuries[6] and continue to intrigue scholars and leaders of varying political orientations.[7] Social power is the ability of one person to influence another person. Influence is the pressure a person, organization, and/or institution exerts on someone else that leads to changes in attitudes, opinions, values, goals, and/or behaviors.[8]

French and Raven[8] identify six different types of power: reward, coercive, legitimate, referent, expert, and informational. A description of these types of power based on Bruins'[7] summary is offered in the following paragraphs.

Coercive and reward power are the top two tiers (force and economic) of the Flora model of social control triangle discussed in Chap. 2. Policing, enforcement, fines, and economic incentives can result in socially appropriate positive and negative responses from those targeted. However, changes are only superficial. People may comply because of fear or economic gain but will need continued surveillance to sustain compliance. They are not likely to change privately held beliefs, attitudes, and values.

Fines must be perceived as of significant cost to deter negative behaviors and economic incentives must be viewed as having high reward value or they lose their power. Social and political counter-responses to the use of perceived inappropriate force and economic incentives can result in strained relationships, blocking actions that lead to polarization, and loss of trust in government and economic institutions. Social responses to "power over" structures include advocacy and conflict-based

---

[6]Machiavelli, Niccolo 1532, a Florentine, Italy public servant and political theorist who wrote *Principe (The Prince)* a political treatise; Hobbes, T. 1651. *Leviathan*, edited by Oakshott, M 1962. London: Collier-Macmillan; Hunter, F. 1953. "Community Power Structure." In *The Search for Community Power 1974*, edited by W.D. Hawley and F.W. Wirt. Englewood Cliffs, NJ: Prentice-Hall.

[7]Bruins, J. 1999. "Social Power and Influence Tactics: A Theoretical Introduction." *Journal of Social Issues* 55(1):7–14.

[8]French, J.R.P. Jr. and B. Raven. 1968. "The Bases of Social Power." In *Group Dynamics*, pp. 259–269, edited by D. Cartwright and A. Zander. New York: Harper & Row.

social actions[9] as well as polarized positions that become barriers to effective public problem solving.

Legitimate power rests on beliefs that a person or an organization has a legitimate right to exert influence and the recipient has an obligation to accept this influence. For example, when a citizen of a community or country accepts the democratic political process as legitimate they will also accept their obligation to comply with the results of elections and laws that are passed even if they did not support them prior to the collective vote. Similarly, when members of an organization elect a chairperson, members accept the chair's right to expect members to comply with chair decisions because they view that leadership as legitimate. Legitimate power leads to private acceptance and does not require surveillance to ensure compliance because the individual has internalized the right of that person to exert pressure.

Referent power is based in self-identification with the person attempting to exert influence. Neighbor-to-neighbor and peer-to-peer influence occurs because the person exposed to the influence of the respected peer acknowledges common goals and recognizes similarities to his/her own situation. For example, farmers identify with other farmers whom they perceive as having similar goals and experiences. These mutual understandings reduce relationship barriers and provide a basis for influencing each others' beliefs and attitudes toward adopting (or not) certain practices. This self-identification leads to a personal acceptance and to internal changes that result in behavior changes. The replication of a conservation ethic so that it becomes a community culture can be the result of referent power, with others following the example of someone considered a successful conservation farmer.

Expert power requires that the person being influenced accept and value the superior knowledge or experience of the organization or person attempting to induce change. Trust in an organization or a person's credentials leads to compliance when expert power is exerted. Scientists, educators, and technical professionals of universities and public agencies have expertise that is needed to manage and protect the watershed. When EPA, state departments of natural resources, or university scientists share data from water monitoring or scientific studies and interpret the results, they are displaying expert power. While individual farmers may not like these findings, they will accept them if they acknowledge the expertise. When a public agency combines their expert power with coercive power, they can exert very strong social pressure. Under voluntary conditions (no coercive power present), this expert knowledge has no power to change individual behaviors unless the expert individual or institution is perceived as a credible and trusted source of information. Thus, the relationship between the expert and the potential recipient of the expertise is critical if beliefs and behaviors are to change.

Informational power has the direct capacity to transform attitudes, beliefs, opinions, and behaviors, creating lasting change without relationship building. Those who seek

---

[9]Minkler, M. 1999. *Community Organizing and Community Building for Health*. New Brunswick, NJ: Rutgers University Press.

information and learn how to apply it to their own personal situation create new mental viewpoints that are used to guide future behaviors. Traditional printed materials (e.g., newspapers, newsletters, magazines, and extension bulletins) as well as electronic media (e.g., TV, radio, DVD, email, Internet, Twitter, and other social media networks) are sources of information and therefore are potential sources of influence and power. However, not all information will be perceived of equal value or trustworthiness. Perceived relevance and validity of the information are what gives information the power to change how people think and act.

Prestige and Status

Power emerges from multiple sources including structural position, personal charisma, expertise, and serendipitous opportunity. Complex networks based in both enduring social relationships and transactional relationships affect the capacity of an agency or community group to get a high proportion of watershed residents to embrace proposed changes in watershed management. Enduring social relationships are reoccurring exchanges over time between particular people (e.g., people living near each other – communities of place) with mutual dependencies and may or may not have an explicit purpose (e.g., protecting a shared watershed).[10] Transactional exchanges are described in terms of costs and benefits (financial, emotional, gifting, and obligations) to each participant in the exchange and often involve organizations and institutions outside of personal, close relationships.

The power and prestige structure that occurs when people are oriented toward the accomplishment of a collective task is based in the expectations of what each member will contribute to completing the task.[11] Those with perceived lower status will have limited ability to influence action because they will have lower performance expectations and as a result will be given fewer opportunities to speak or contribute or will have their contributions evaluated poorly or ignored. Conversely those of perceived high status will be able to not only persuade but also exercise dominant, directive power over group members.[11]

Status beliefs about people are formed from reoccurring personal encounters. "When people who differ on a socially recognized characteristic interact in regard to a shared goal, a status hierarchy will emerge among them…."[12] In an agricultural watershed community, socially recognized characteristics might be: frequent adopter of conservation practices, likely to risk innovation, has a lot of knowledge about agriculture technology, or generous to neighbors and likely to help others.

---

[10] Cook, K.S., C. Cheshire, and A. Gerbasi. 2006. "Power, Dependence, and Social Exchange." Chapter 9. In *Contemporary Social Psychological Theories,* edited by P.J. Burke. Stanford, CA: Stanford University Press.

[11] Correll, S.J. and C.L. Ridgeway. 2006. "Expectation States Theory" Chapter 2. In *Handbook of Social Psychology,* edited by J. Delamater. New York: Springer Science+Business Media, LLC.

[12] Correll, S.J. and C.L. Ridgeway. 2006. "Expectation States Theory" Chapter 2. In *Handbook of Social Psychology,* p. 44, edited by J. Delamater. New York: Springer Science+Business Media, LLC.

If this fledgling status belief is reinforced in future encounters, it becomes stable and the person is viewed from the perspective of what "most people" think. However, even a slight challenge to social consensus can weaken status beliefs and disrupt the emergence of new ones.[13] Status beliefs have implications for how reputations are created and why some people and organizations are held in higher regard and have the power of influence and others have less.

One theory that attempts to explain the factors determining power and how it is used is called exchange theory. Differences in power are derived from the degree of dependence people have on each other for resources of value and of what they are able to offer in exchange.[10] Power-dependency relationships are motivated by a desire to increase gains and avoid losses. They are influenced by: (1) the mutual dependence of interacting parties based on the value of the resources to be exchanged and availability of alternative sources, (2) the intensity of recurrent exchanges over time, and (3) the point at which valued outcomes have reached diminishing usefulness.[10]

Individual Influence, Changes of Mind, and Willingness to Act

Enduring Relationships

Friends, neighbors, family, and community social acquaintances are all sources of information exchange and social pressure. Each exerts influence and power over others and affect whether and to what extent water and other environmental problems are identified. Further, they affect what solutions are considered and decisions about how they are implemented.

In 2008 and 2009, the Iowa Learning Farm conducted 15 listening sessions with Iowa Department of Natural Resource (IDNR) conservation officers, Natural Resource Conservation Service (NRCS) professionals, and conservation-minded farmers.[14] These agency technical professionals offered valuable insights on the impacts of enduring relationships and how individuals influence each other. One recalls a situation where family relations made the difference:

> ...I was at a farm where the grandson was there with the grandfather, and the grandfather was cooking lunch for them. We were trying to get permission to drill [for soil cores]....
> I really got the sense that the grandfather wasn't so sure about it, and the grandson is like, this is a good thing. And then I ended up talking to the son and the father of the

---

[13]Ridgeway, C.L. and S.J. Correll. 2006. "Consensus and the Creation of Status Beliefs." *Social Forces* 85(1):431–453.

[14]The goal of these 15 listening sessions conducted by Iowa Learning Farm with Iowa NRCS, IDNR, ISU Extension, and conservation-minded farmers was to better understand how these professionals perceived the issues affecting conservation, their roles as conservationists, and their ability to influence change.

grandson and they were like, 'well, you can drill holes anywhere on our farm' ... so I think you do see with some of those older farmers that ... they really do value the opinions of their children and their grandchildren.... [I was] fortunate enough to have the younger person there who sees some value in knowing the geology of their land ... [because] all of us run into [the attitude] we don't want the government to know anything about our land (0129090406).

A NRCS professional suggests that to be effective he needs to build a relationship with farmers before he can use his expert influence to educate:

We have lots of programs that we can use. I think we need to get back to our basics and get back out to the field and start doing handshaking and kitchen table talking and start educating the landowners as to what is out there and what we have. I guarantee that'll sell more programs faster with that type of approach. We can throw money at the problem all day long. They're never going to fix it because they [farmers] don't understand it. I'm doing it because I'm getting a payment. When the payment's gone ... it didn't work [and the farmer stops the practice].... So we need to get back out at the kitchen tables and say, this is what we've got, this is where we need to go, this is how you should do it (041008).

The combination of coercive power and expert role often is a deterrent to farmers' sustained behavior change because internal beliefs and opinions have not changed. Several IDNR conservation officers talk about the tensions between the power their agency has to monitor and enforce compliance and their desire to influence farmers to voluntarily change their practices. Some believe that perceptions of the IDNR role as a monitoring and enforcement agency creates a lack of trust that is a barrier to their ability to influence farmers' personal environmental management goals. One conservation officer says it is:

...a respect issue, and I think that's huge. I mean, they're going to listen more to another farmer who has implemented conservation practices than they are to us in the DNR, because ... [we have] more adversarial relationships [because of our role] (0129090421).

The tension between the agency role and a desire to be effective in working with farmers is echoed by a NRCS professional:

I think some people just have a fear of the government, which is the stigma around.... I mean, they're going to trust their neighbor who works at the co-op before they want to walk into a government building. I think there's some people that are like that (052008).

Other IDNR conservationists observe:

...you have better success with ... farmers that are in it voluntarily or they're in it to get the money to do it ... [they] try to persuade their neighbors or friends ... they don't want us out there because we're regulatory and when we see problems, you don't have any choice but to fix it, our job is to require you to fix it ... (0129090421).

Agencies' priorities [should be] to spend more money and time and personnel in direct contacts with the landowners, because that is what changes things ... our own agency talks a good game ... but doesn't always put the real funding towards that kind of thing, when we may spend a couple hundred thousand or a couple million on something that is pretty indirect for really helping the landscape ... (0129090424).

Neighbor-to-neighbor relationships and local leadership are perceived as effective in increasing conservation program participation and getting good practices implemented. Several agency listening participants agreed they needed to find community and agricultural leaders and those leaders are most effective when they have a long-term vision.

> They've got to be able to lead other people into programs … if it's peer pressure, then if you can get those people who are in those leadership roles in their ag community or farm community, then hopefully other folks would follow suit (0129090115).

Peer-to-Peer Influence

The influence that individuals have on each other as they interact on a regular basis is illustrated by a 2006 key informant survey of 360 conservation-minded Iowa farmers in 75 randomly selected Hydrological Unit Code (HUC) 12[15] watersheds.[16] In this research, the average farmer weekly talks to at least eight other farmers and almost nine friends and neighbors that do not farm. They also belong to, on average, approximately six groups and organizations such as Farm Bureau, Corn Growers Association, Cattlemen Association, Pheasants Forever, church, and community service groups.[16]

Friends and neighbors have significant influence on farmer respondent satisfaction with their current conservation measures and willingness to change practices. Farmers who report talking with more farmers weekly are more likely to say they are satisfied with their conservation efforts compared with farmers who talked to fewer farmers. Further, the more nonfarm neighbors and friends farmers report talking with weekly, the more likely those farmers report dissatisfaction with their conservation efforts. In addition, the more social organizations a farmer belongs to, the more likely they are to not be satisfied with their conservation efforts. Thus, in this study farmers, reinforce each others' satisfaction with their current conservation efforts, while contact with others in the community reinforces the expectation that they could do more.

---

[15] HUC is the acronym for Hydrologic Unit Code. Every hydrologic unit is identified by a unique HUC consisting of 2–12 digits based on the levels of classification in the hydrologic unit system. A hydrologic unit describes the *area of land* upstream from a specific point on the stream (generally the mouth or outlet) that contributes surface water runoff directly to this outlet point. Another term for this concept is drainage area. It is delineated by starting at a designated outlet point (usually the river mouth) and proceeding to follow the highest elevation of land that divides the direction of surface water flow (usually referred to as the ridge line). Hydrologic Unit Codes (HUC) data describe watersheds as polygons, defined by digital elevation model data. HUC basins decrease in size with an increase in levels. For example, HUC6 watersheds are major river basins while HUC12 watersheds are 10,000–40,000 acres or 15–62 mi^2.

[16] Morton, L.W. and C.Y. Weng. 2009. "Getting to Better Water Quality Outcomes: The Promise & Challenge of the Citizen Effect." *Agriculture and Human Values*. In special issue on *Civic Engagement and Alternative Rural Development* 26 (1):83–94.

Organizational Power and Influence

The collective efforts of individuals are often more powerful than a single individual to influence and change behaviors. The formation of watershed groups, alliances, and councils leverages the different types of power that each individual brings to the group. Member connections and relationships outside the group become a resource that the group can use to achieve their purposes. Within these groups, trust relationships are built and members begin to self-identify even with others who represent different viewpoints and, as a result, they are able to direct their social power toward a common goal. Watershed groups that utilize the power of information, including learning how the water–land ecosystem works and new technologies, can be particularly effective in changing their own and others' internal beliefs and in turn increase conservation-associated behaviors.

The Applegate Partnership,[17] a community-based nonprofit organization involving industry, conservation groups, natural resource agencies, and residents, is an example of group power addressing environmental concerns. Created in 1992 as a collaborative effort to maintain the long-term health of their watershed and stabilize their local economy, the Applegate Partnership was formed to overcome high levels of distrust and conflict and implement landscape-level planning.[18] Years of constant conflict and battles pitting logging interests against environmental concerns on US Bureau of Land Management and US Forest Service lands had polarized the community and stymied its ability to address either sector's issues. The Applegate Partnership was able to move beyond polarized conditions by creating referent or self-identity power through new trust relationships and a collective sense of purpose among citizens. Seven years after its establishment, interviews with 20 partnership members affirm that all have higher levels of trust – even with those previously viewed as adversaries. A federal official says there has been an increase in understanding of each other's issues and more shared respect despite differences.

Organizations that hope to wield power and influence must build trust among their members and externally with others. Weber[18] notes that, although the Applegate Partnership is a formal institution, trust and influence are individual based. He finds that "the openness and iterative deliberations of the partnership make it easier to discern who is worth trusting and who is not, with a premium placed on forthrightness, integrity, and honesty."[19] Forthrightness, integrity, and honesty are social characteristics that the group value, rewarding those who demonstrate them and censuring those who do not by diminishing the influence they have.

---

[17] http://www.roguebasinwatersheds.org/SectionIndex.asp?SectionID=3.

[18] Weber, E.P. 2003 "Bringing Society Back." In: *Grassroots Ecosystem Management, Accountability, and Sustainable Communities*. Cambridge, MA: MIT.

[19] Weber, E. P. 2003. "Bringing Society Back." In: *Grassroots Ecosystem Management, Accountability, and Sustainable Communities*, p. 137. Cambridge, MA: MIT.

The power of the collective group to influence member attitudes and actions is demonstrated by 10 out of 20 Partnership interviewees claiming that, as a result of participating in the Partnership, they are more willing to consider the effect of a proposed partnership decision on the outside world (rather than only the effect on themselves) as a starting point. Further, the Partnership has increased the proportion of members from 30 to 45% who say they are willing to give more weight to watershed community benefits versus their personal self-interest. The act of participating in a group with watershed-wide goals appears to have influenced members to shift their personal perspectives and increased their willingness to seek solutions for the whole watershed.

Information Power

The adage "information is power" is demonstrated by the millions of advertising dollars private industries and public groups have invested to get people's attention and induce them to purchase a product or do something different. Most institutions utilize a variety of media to exert information power. Information and education are essential if citizens and watershed groups are to influence other citizens and community sectors that are not knowledgeable or friendly toward watershed protection goals and management strategies.

A 2006 four-state survey of the Heartland Region reveals that media and personal contacts are primary sources of influence that lead to changes of mind on environmental issues.[20] In a stratified random sample of all residents in Iowa, Missouri, Kansas, and Nebraska, almost 56% reported changing their minds based on news coverage. Fifty percent said first-hand observation and 41% said conversations with other people were primary reasons for changing their mind on environmental issues. The media, newspapers (72%) and television (61%), dominated public institutional sources of water quality information. Use of the Internet for information was strongly skewed by age. Overall, approximately one-third of respondents reported they would visit a web site for information and tips on water quality issues. However, 54% of those younger than age 40 years compared with 22% of those aged 60–70 years old said they are likely to visit a web site for water quality information.

Analyzing for Influence: A Political Accounting System

Local leaders, community development specialists, extension educators, and other watershed professionals can make use of existing sources of power and influence to move watershed plans forward. The challenge is to identify key players and

[20]Morton, L.W. and S. Brown. 2007. *Water Issues in the Four State Heartland Region: A Survey of Public Perceptions and Attitudes about Water.* The Heartland Regional Water Coordination Initiative Bulletin #SP289 Iowa State University Extension. http://www.heartlandwq.iastate.edu.

social networks of power and influence that are currently sources and barriers to change. An understanding of the political and social connections among people offers important clues to interventions to achieve water quality goals. Coplin and O'Leary[4] propose a political accounting system for analyzing power and influence and organizing the large amount of data and facts needed to solve a political problem. Box 4.1 below summarizes the main elements of this accounting system applied to a collective effort at watershed management.

The first two elements in this accounting system are familiar requirements for local group formation and the writing of technical watershed management plans. The group must have some level of agreement on what is important and what they want to accomplish by working together. This clarity is necessary so the group knows toward what goals they wish to direct their and others' power and influence. Genskow and Prokopy[21] identify NPS management behaviors and actions that watershed groups frequently specify to get to their goals. They include: (1) increasing awareness of watershed issues, (2) changing attitudes so they are supportive of NPS management actions among target audience, (3) reducing constraints for using appropriate practices, (4) increasing capacity to address NPS management issues in the project area, and (5) increasing adoption of NPS management practices by a target audience.

The other four elements (position, salience, power, and friendship/hostility affiliation) of this accounting system are information to be collected about social relationships and the nature and strength of those connections. These data provide

---

Box 4.1 Adaptation of Coplin and O'Leary political accounting system (1976)

1. Define the problem in terms of a desired concrete outcome. The group needs a *vision* and *concrete goals* to know where they want to go and to be able to tell when they arrive.
2. Specify the kind of *behaviors and actions* that need to happen to get to the goal.
3. *Position*. List key people, groups and institutions that affect water quality outcomes; identify their priority issues.
4. *Salience*. Make a best estimate of how each person, institution, and group will respond to the water goals and proposed actions (issue position) and the strength and importance they attach to the issue (salience) using a negative to positive scale [−10…0…+10].
5. *Power*. Identify the power each person, institution, and group has to block or make the group's goals happen [0–10]. A lot of zeros suggest you don't have the right people identified as power players.
6. Explore the affiliation (hostility-friendships relationships) among key players; what is their history of how they support and align with each other.

---

[21]Genskow K. and L.S. Prokopy. 2008. The Social Indicator Planning and Evaluation System (SIPES) for Nonpoint Source Management: A Handbook for Projects in USEPA Region 5. Great Lakes Regional Water Program. Publication Number: GLRWP-08-SI01.

the basis for specifying strategies that have the potential to shift power and influence toward the goals of the group. This information can be gathered by direct and indirect conversations, past experiences, observations, letters to newspapers, and public dialogs.

A list of key players who are thought to have power to help or stop the watershed group from accomplishing their goal is the first set of data needed. Some key players will have a publicly declared a position, others may only allude to their opinions. Key players are identified by their power and relationships, not necessarily by their position on the issue. Often it is only an educated guess or best judgment of where key players stand positionally.

Next, each key player is analyzed based on their perceived position on the watershed goals, how important they think they are (salience), the power they have over others to get them to support or undermine the watershed group efforts, and the friendship–hostility relations each player has with others. Salience (how important the issue is to the key player) is different from their position on a watershed issue. For example, the key player may publicly say they think reducing phosphorous levels is an important goal of the watershed group. This is their position. However, when asked to do something on their land to reduce phosphorous and they respond that they do not have the resources or the time to work on it right now, they are signaling that it is not a very high priority and has a low salience.

Once this analysis is complete, you will have a web of influence structure that can be used to develop strategies to shift the influence people have on your goal. Intervention strategies can range from finding ways to change the position of one or more persons on the issue to increasing the salience or importance of the issue, increasing the power of those who agree with you (or weaken the power of those who oppose you), and make friends and win over your enemies.[4]

An important intervention for watershed groups is to try to increase the perceptions of importance or *salience* of the water quality issue for those they identify as most likely to support actions. To do this, the targeted individuals have to be addressed in terms of their core beliefs. Sabatier and Jenkins-Smith[22] find that core beliefs are filters to making sense of new information and taking actions. Thus to increase salience on watershed issues, individuals' normative values and core beliefs about stewardship and conservation must be identified and the watershed group needs to offer new information that is compatible with existing deep core beliefs. This new information strengthens the need to act. When scientific and technical information are at odds with core beliefs, as often happens in government designations of impaired water bodies, an external crisis may be the motivator to act.[23] This is why the threat of coercion or force (e.g., EPA regulation and fines) can increase awareness of consequences of inaction and has the potential of positive action to make a difference.

[22]Sabatier, P.A. and H.C. Jenkins-Smith. 1999. "The Advocacy Coalition Framework: An Assessment." In *Theories of the Policy Process,* edited by P. Sabatier. Boulder, CO: Westview.

[23]Weaver, M. and R. Moore. 2004. "Generating and Sustaining Collaborative Decision-making in Watershed Groups." Presented at 67th annual meeting of the Rural Sociological Society Sacramento, August 11–15.

Key leaders in a watershed can influence others to see the importance of the watershed goal by championing the vision and modeling concrete actions for change. Peer-to-peer influence among farmers is one strategy for increasing management practices likely to improve water quality. A 2007 survey of 1,100 agricultural landowners in the 12 subwatersheds of the Lower Big Sioux River (Iowa)[24] reveals significant influence of peer practices on other landowners in utilizing practices such as soil testing, integrated pest management, systematic crop scouting, farm based records, nutrient management, and manure structures to reduce water pollution. Peer influence is also evident, although to a lesser degree, in using reduced tillage, no-till, contour strip farming, and installation of terraces.

Few people formally engage in the Coplin and O'Leary political calculations to discover who will influence what outcomes, just as few people actually do full-scale benefit/cost analyses.[25] The real point is to systematically develop the discipline of thinking in terms of position, power, salience, and friendship–hostility relationships and the tradeoffs that occur.[25] Coplin and O'Leary[26] suggest that a variety of patterns of compromise may also give the group power to achieve their goal. These include: (1) delay or postpone contentious decisions, (2) give each key person a portion of what they want, and (3) create an atmosphere of trust and compromise. They also recommend offering extra incentives to those with higher salience (those who give the issue a high priority) because these are the people most likely to act in positive or negative ways toward your goal.[27]

A Balancing Act

Grassroots community and watershed-based approaches to aquatic resource management have a difficult balancing act between local landowner control and creating the capacity of citizens to learn and benefit from government technical and scientific support while minimizing bureaucratic risks and costs.[28] There is a constant tension between the dominant power structure, which has financial and technical resources, and the effort to create and maintain an open participatory process among local community members.[23] The very nature of watershed partnerships that are diverse can advantage major stakeholders and replicate established, recognized

[24] Lower Big Sioux River Watershed Survey Nov 2007; 4,439 surveys mailed by 3 county Soil & Water Conservation District offices to all landowners. Single mailing, no follow-up. $N=1,110$ completed surveys 25.2% response rate.

[25] Filipovitch, A.J. 2005. PRINCE Analysis. (http://Krypton.mnsu.edu/~tony/courses/609/Frame/PRINCE.html) Retrieved April 6, 2008.

[26] Coplin, W.D. and M.K. O'Leary. 1972. *Everyman's Prince: A Guide to Understanding Your Political Problems*, pp. 168–170. North Scituate, MA: Duxbury.

[27] Coplin, W.D. and M.K. O'Leary. 1972. *Everyman's Prince: A Guide to Understanding Your Political Problems*, p. 43. North Scituate, MA: Duxbury.

[28] Habron, G. 2003. "Role of Adaptive Management of Watershed Councils." *Environmental Management* 31(1):29–41.

interests to the disadvantage of those citizens who are not organized, not informed, and lack resources.[23] Thus a watershed partnership risks becoming transactional – negotiating compromises; rather than transformational – generating new ideas and reconstructing underlying beliefs and interests.[23] Weber[18] finds that transformation can occur in diverse partnerships when trust relationships are built and the focus is accountability to a broad cross-section of society. Thus diverse watershed partnerships can replicate the power and influence of established interests, or they can transform the underlying social will to improve if trust relationships are built and the focus is accountability in the public interest.

Traditional concepts of power are hierarchical. They stress individual and group "power over" other individuals and groups. This kind of power may lead to compliance but often does not change underlying internal beliefs and attitudes that sustain personal actions. Changes in actions occur because of the force of unequal power relations.[3] Some people control more valued resources than others. This leads to unequal power relationships as social debts are incurred and in turn paid for by acts of compliance.[10]

When power relations are unequal, the less powerful will attempt to find strategies to equalize or obtain dominance. Polarization is often the result. Polarization is the extent to which there are two distinct positions or viewpoints and few or no persons who are friendly with both views.[4] Polarization is a blocking tool if one group does not want consensus or action to occur. Its practice can shift power and is often used by a less influential person or group as an attempt to gain power over a dominant, more powerful group. Issues can be depolarized if friends and kin are mixed on both sides of the issue, thereby increasing problem-solving potential. Polarization can also be minimized when the major issues that could create polarization are identified early and the people who can bridge extreme positions are motivated to use their influence and power.

Another way to avoid polarization is to use a collaborative or partnership approach that applies a shared power form of social pressure rather than "power over" to achieve goals. Grass roots environmental management processes incorporate shared power strategies in conjunction with the power of internal personal beliefs and knowledge to encourage learning about natural systems, collaboration, and community capacity building.[29] The practice of community development is a shared power strategy that builds citizen competencies and leadership while empowering social changes that are concrete, pragmatic, and action specific.[9]

Influence is socially induced changes in beliefs, attitudes, and expectations without the force of sanctions.[3] Influence is primarily persuasion, information, and nonbinding advice.[3] Sources of influence are experts, media, friends and neighbors, organizations, and institutions. Neighbor-to-neighbor or referent power has been

[29] Weber, E.P. 2003. "Bringing Society Back." In: *Grassroots Ecosystem Management, Accountability, and Sustainable Communities*. Cambridge, MA: MIT; Morton, L.W. 2008. "The Role of Civic Structure in Achieving Performance-based Watershed Management." *Society & Natural Resources* 21(9):751–766.

shown to be a particularly strong motivator of behavior change and of deep, long-lasting changes in beliefs and attitudes. Community leaders and watershed specialists who want to make a difference in their watershed will begin with building trust relationships and providing access to information. These connections will give them power to influence others and become effective champions of change.

Chapter 5
Turning Conflict into Citizen Participation and Power

Jeff Zacharakis

The voice of watershed residents – whether in agreement or disagreement – is an essential part of the learning and negotiation process needed for internal change in beliefs and active engagement, which is the only route to sustainable adoption of appropriate conservation practices. While conflict is generally thought to be a barrier to community action, it can be used as an asset to strengthen local watershed councils and other community organizations. The underlying principle of local control and local leadership is central to leveraging conflict to achieve positive community watershed outcomes. During my 11 years as an extension community development specialist, I found that even when local leadership is inexperienced, under conditions of local control, conflict assumes a different character than when state and federal agencies attempt to drive change from the top down. In this chapter, I will discuss conflict and authority's political dimensions, types of conflict, the importance of turmoil and conflict, and strategies to enhance the value of conflict as a community development asset. My community development work in Iowa's Maquoketa watershed is used as a case study to illustrate these concepts.

Conflict's Political Dimensions

The word politics usually conjures up images of government representatives telling us over the radio and on TV how to view a particular event or situation. Yet the historical definition of politics includes everyday citizens working through difficult issues such as water pollution, watershed preservation, and restoration. Under these circumstances, water becomes the focal point of local politics. Watersheds are not just geographic units, they are also political entities where special interests are negotiated and power is exercised.

J. Zacharakis (✉)
Department of Educational Leadership, Kansas State University, 326 Bluemont,
Manhattan KS 66506, USA
e-mail: jzachara@ksu.edu

L.W. Morton and S.S. Brown (eds.), *Pathways for Getting to Better Water Quality: The Citizen Effect*, DOI 10.1007/978-1-4419-7282-8_5,
© Springer Science+Business Media, LLC 2011

The word politics comes from the Greek word "polis," meaning state or community as a whole. Plato,[1] in his essay, *The Republic*, describes the ideal state and how politics is a means to achieve it. Aristotle[2] writes in *The Politics* that "Man [sic] is by nature a political animal." Engaging in politics is fundamental to human nature. Residents in a watershed will inherently seek to preserve their own interests, ideas, and preferences and in the process produce multiple perspectives. *The Blackwell Encyclopedia of Political Thought* elaborates, "Politics presupposes a diversity of view, if not about ultimate aims, at least (about) the best ways of achieving them."[3] When citizens meet to discuss a community issue, opposing interests and conflicts are bound to occur. The negotiation of a common agenda or finding common ground so decisions can be made is a political act.

Politics is a social response to manage competing special interests and a means to create a more organized and peaceful society. In a democracy, it provides methods to resolve conflict through civil discussion and rational compromise. The goal is to prevent or reduce the chaos and potential disintegration of society that could otherwise result. Leftwich[4] writes that "politics comprises all the activities of co-operation and conflict, within and between societies" and all of the areas of life – social and biological – wherever humans are involved. Watershed management is not exempt, conflict and politics exist because there are opposing goals and competing cultures and ideologies.

The resolution of conflicting opinions requires negotiation and agreement among affected parties. Three elements of politics are necessary to resolve social disagreements, "persuasion, bargaining and a mechanism for reaching a final decision."[3] This process can be facilitated by persons who are able to bridge differences of opinions, thereby mediating the public discussion in a search for solutions. In the next sections, I describe the Maquoketa River Watershed, types of conflicts that occurred, and the water politics that were used to effectively build a common agenda and actions to improve local waters.

Maquoketa River Watershed

The Maquoketa River Watershed in northeast Iowa is one of the largest contributors of excess sediment and nutrients to the upper Mississippi River. More than 61,000 people live in the 1,879 square mile Maquoketa basin, where land use is primarily

---

[1] Plato. 1987. *The Republic*. London: Penguin.

[2] Aristotle. 1996. *The Politics and the Constitution of Athens*, p. 1. Cambridge: Cambridge University Press.

[3] Miller, D. 1987. *The Blackwell Encyclopedia of Political Thought*, p. 390. Oxford: Basil Blackwell.

[4] Leftwich, A. 1984. *What is Politics?*, pp. 64–65. Oxford: Basil Blackwell.

agricultural. It is a picturesque area, characterized by small rural communities and small- and medium-sized family farms situated in rolling hills with highly fertile soil.

Throughout the United States, nonpoint source pollution has been identified as the leading cause of water quality degradation, most of which is attributed to agricultural practices.[5] In 1998, Iowa established the Maquoketa Watershed Project as a multi-agency initiative to deal comprehensively with nonpoint source pollution. The project included an effort to promote citizen-led watershed councils in each of the Maquoketa's 25 Hydrological Unit Code (HUC) 11-digit (HUC 11) sub-watersheds. The intent was to strengthen citizen awareness and local participation by developing a comprehensive plan to address its environmental problems. A contribution to this effort was a 1999 grant from the US Environmental Protection Agency (EPA) Region 7 to Iowa State University Extension "to develop local leadership with long-term vision and commitment to deal proactively with nonpoint source pollution issues."[6]

In the Maquoketa basin, sub-watersheds designated HUC 11 by the Natural Resource Conservation Service (NRCS) range in size between 30,000 and 50,000 acres, with between 1,000 and 5,000 residents. Focusing community efforts at this level provided a management tool for involving more citizens in watershed decision making. Larger regional units favor authoritative decision making and control by government agencies and fewer citizen representatives. Management often consists of generalized solutions developed for the entire watershed basin. The devolution to smaller units provides opportunities for more direct citizen involvement. However, it must be recognized that increasing citizen participation creates a corresponding increase in potential conflicts than if the large watershed was managed as one unit.

The Maquoketa basin sub-watershed environments are very diverse. For example, in some areas of the Maquoketa basin, the primary agricultural enterprise is a corn–soybean rotation with little livestock. In other areas, there are concentrations of family dairies, a predominance of hog production, or clusters of family feedlots where beef is finished for market. These variations are the result of soil types, topography, and microclimates as well as historical settlement patterns, cultural traditions, and religious and ethnic concentrations. Cultural heritage and religious affiliation serve to facilitate social gatherings where information is exchanged and friendship–hostility relationships are reaffirmed.

---

[5] Shepard, R. 1999. Making Our Nonpoint Source Pollution Education Programs Effective. *Journal of Extension* 37(5). (http://www.joe.org/joe/1999october/a2.html); Schilling, K. E. and C. F. Wolter. 2001. Contribution of Base Flow to Nonpoint Source Pollution Loads in an Agricultural Watershed. *Groundwater* 39(1):49–58.

[6] Maquoketa Quarterly Reports. 1999. *EPA Region VII Water Quality Cooperative Agreement*, p. 1. Iowa State University, Ames, IA (October and December).

Regionally, a farmer may be known as Joe Smith who farms 1,300 acres of corn and soybeans, but, within the sub-watershed, citizens know each other intimately – neighbors are likely to know how many generations the farm has been in the family, which kids are in college, and many personal details. More importantly, neighbors know who is a "good farmer" and who is not. However, the definition of "good" varies considerably from sub-watershed to sub-watershed, depending on the cultural, economic, and social expectations.

Some of the sub-watersheds have sizable towns within their boundaries, ranging in population between 1,000 and 5,000 people. These urban sub-watersheds also have unique geological situations, cultures, and specific water quality problems that make it difficult to group all of them together and develop single solutions. Their city councils work differently and have unique relationships to the watershed that are different from county commissioners or township trustees. One community may lie on heavy clay with relatively tight soils, another may have been built on a limestone formation with highly porous and fracture hydrology, creating easier movement of pollutants. Another may have a meat packing plant that uses and discharges large amounts of water. A single approach to dealing with nonpoint source problems in all the urban communities in the Maquoketa River watershed would be as if all the farms were grouped into one large management district rather than multiple Soil and Water Conservation Districts. A closer look at the social relations in the Maquoketa River basin offers an illustration of how ecological and social complexities affect the kinds of conflicts that occur and strategies for resolution.

Types of Conflict in the Maquoketa Watershed

Examples of conflict found in three sub-watershed projects in the Maquoketa River Basin are explored in this section. These projects were selected because they are relatively complex and controversial. Instead of trying to manage and minimize conflict, the conflict became an asset that enhanced project outcomes. Potentially explosive variants of conflict within these projects were managed while maintaining the vitality and energy that conflict brings to a project.

Overlaying all levels of conflict described in this section, is the power of government and the fear of its regulation over agriculture, which proved to be the primary reason why farmers initially became involved in these watershed management projects. Between 1999 and 2000, farmers throughout the region became aware that the EPA was proposing regulation of all farming operations in the same way that industries were regulated to reduce nonpoint source pollution. One threat was the possibility that livestock operations with more than 300 animal units would fall under EPA regulations instead of the existing Concentrated Animal Feeding Operation threshold of 1,000 animal units. Farmers were nervous and upset. Fear that government would tell them how to farm made many feel like victims of unreasonable blame for the watershed's environmental impairment. Further, they had few or no opportunities to discuss, understand, and accept the environmental watershed assessment conducted by state and federal agencies that gave rise to the "impaired" designation.

As the result of their concerns, several community members in each of the sub-watersheds requested assistance from Iowa State University Extension to work with local leaders and other state and federal agencies to organize community forums to discuss specific issues and opportunities. While the sub-watersheds differed in character and their priorities, they shared the same goal of forming a local water-shed council. In each of these watersheds, I (in the role of the extension community development specialist) worked closely with a few local leaders, such as Soil Conservation District Commissioners and respected farmers, to form a steering committee. This steering committee developed a strategy to invite everyone who lived in the watershed to the public forum. The number one responsibility of the steering committee was to ensure that every resident in the watershed knew that they were not only welcome to attend the forum, but that their participation was essential to the future success of the watershed council. Because of the small popu-lation residing in these watersheds, the steering committee played a crucial role in overcoming any potential conflict that might arise. Through their local leadership, people felt welcomed and safe when attending these meetings. Large numbers of residents – 100–150 people – attended each of the community forums in the three sub-watersheds.

Types of Conflict and Turmoil

Within the three watershed projects, many types of conflict and turmoil emerged. The list below was developed while working in these watershed projects. Their experiences can be applied to any complex project.

Family conflict can strongly impact the potential success of a project if it is not taken into consideration when developing strategies. Most farming operations include parents, brothers and sisters, and aunts or uncles. In some farming partner-ships, family members who share in the ownership do not live on the farm or par-ticipate in its management. In one instance, a farmer arrived with his brother and father. Even though they shared ownership, his brother lived in another state and was most interested in receiving his rent, and his father was less than 5 years away from retirement and did not want to invest any money into improving their farming practices or upgrading their feedlots (two of which had streams running through them). It was a tremendous victory for this farmer to successfully convince his family to attend these meetings with him even though there were years of conflict between these men on how to plan for the future of the farm.

While individuals in rural communities rarely speak openly about family issues, conflict *between neighbors* is usually known to everyone – except to the commu-nity development specialist who is trying to assist the formation of a watershed group. The specialist may only know that certain neighbors are not on speaking terms, while the rest of the group knows the reason for their discord. Neighbors who have had past disagreements did not come to these watershed meetings with expectations of working together.

When possible, being able to identify potential firestorms was essential to avoiding open conflict, while still enabling all parties to continue participating in the process. In one watershed, the project was allocated some federal money to conduct a pilot test of bio-filters as a low-cost strategy for tertiary cleaning of wastewater from family-owned beef lots. When the owner of the largest feedlot in the watershed volunteered the use of his farm for this pilot project, we learned from other producers that selecting that farm would alienate smaller, less visible producers and therefore dilute the community's interest in an important environmental demonstration initiative. As project facilitator, I maintained a neutral position during these discussions to keep the entire group together and the project moving forward. In this case, the watershed council dealt with the problem by continuing to solicit other volunteers until a suitable farm was found – a farm with a feedlot that was more representative of the average feedlot in the watershed. The watershed council's knowledge of how placing the demonstration on the large feedlot might impact public opinion was never discussed publicly. The watershed council used this information very discreetly to avoid publicizing a potentially disruptive situation.

Rural non-farm and small town residents and farm families have environmental conflicts that cause them to cross paths. Livestock manure is a particular problem – where and when it is spread and control of odor. Farmers also have to run their equipment late at night during harvest and planting, and their equipment tears up the road during the spring and fall. On the other hand, rural non-farm residents do not understand the seasonality, physical stress, and tight profits associated with farming. They may be quick to blame farmers for all the pollution that occurs in the watershed even though small town sewer and personal septic failures are also part of the problem. When blaming behaviors occurred, I, as facilitator, reminded the group that our purpose was not to place blame but to work together to solve a common problem.

Watershed citizens invariably want to do their own monitoring before they can accept assessments by agencies. In one watershed, when council members started testing their creek, it was discovered that a small unincorporated village had connected their septic systems, many years earlier, directly into a drainage tile and this contaminated water was flowing directly into the stream. It became clear after this discovery that everyone shared both the blame and the responsibility to improve the watershed.

It is difficult to keep small town residents involved on watershed councils whose target is nonpoint source pollution. One reason in the Maquoketa watershed was the general consensus that the problems were agriculturally related. Another reason was that all of the government cost-share money was allocated for agriculture, and no funds were available to mitigate the impact of small towns and rural non-farm residents. Other than some limited state money to improve septic systems, there was little or no financial incentive for non-producers to stay involved.

Another type of conflict observed in these watershed projects *is conflict between farmers and the government*. While many farmers prefer to have complete control of their operation, they have become dependent upon government payments to maintain their cash flow. In addition, although many farmers have learned how to work within this relationship, it still creates tension similar to other groups who work or live within cultures of dependency, including some businesses and welfare recipients. Statistically much of the profit that farmers have experienced in the last decade is directly due to government programs such as the Loan Deficiency Programs, the Conservation Reserve Program, and ethanol subsidies. Farmers are quick to agree they would love to farm profitably without government payments, but they are willing to work within these regulatory programs for financial reasons, regardless of their personal feelings toward these programs.

Moreover, these programs give government agencies such as the Iowa Department of Natural Resources, Iowa Department of Agriculture and Land Stewardship, and USDA, the authority to become involved in the daily operations of those farmers who want to participate in their programs. The early success of the Maquoketa River watershed project was due in large part to the threat that EPA was moving toward greater regulation of agriculture, and the perception that this level of regulation would result in greater constraints to their daily operations. This regulatory threat, although not enacted in the late 1990s, actually created that type of conflict that encouraged greater producer interest and participation. However, it still led to apprehension on the part of producers to openly discuss these issues, especially when government representatives attended their meetings.

The final type of conflict experienced while working with citizen-led watershed councils is the *tension between local, state, and federal government agencies*. Most state and federal agencies face budget cuts every year. All of them have become more aggressive in seeking new monies, and work hard to document their programs' positive and measurable impacts. Competition between agencies for limited dollars in part explains why these agencies do not work as closely together as they might. However, this element of conflict can be resolved as a win–win situation when different roles and potential contributions to successful outcomes are accounted for.

Iowa State University Extension agricultural educators in the Maquoketa River Basin were primarily involved with producers to improve their manure and fertilizer nutrient management. Extension's community development specialists assisted in developing watershed councils by seeking to empower their clients to engage in critical thought, careful planning, and involvement in democratic decision making and action. Community development theory "promotes broad-based, participatory decision making to initiate social action processes to improve local economic, social, cultural, or environmental situations."[7]

Technical and regulatory agencies have top-down mandates.[8] Their local representatives do not make the rules, and often have little latitude to interpret these rules.

---

[7]Christenson, J. A. and J. W. Robinson. 1989. *Community Development in Perspective*, p. 14. Ames, IA: Iowa State University Press.

[8]Zacharakis, J., L. W. Morton, and J. Rodecap. 2002. "Citizen-led Watershed Projects: Participatory Research and Environmental Adult Learning along Iowa's Maquoketa River." *Adult Learning* 13(2):19–23.

Funding for the same type of watershed work and environmental practices that comes from different agencies may even come with different rules. For the extension specialist, the challenge is to maintain a strong working relationship with federal and state partners while encouraging local residents to mobilize around issues that concern them. The requirements of their differing roles – educator, advocate, technical specialist, and regulator – can create conflict for agency professionals, and certainly may have a negative impact on their clients' involvement in watershed projects.

The Importance of Conflict and Turmoil

The complexity of multiple sources of conflict increases the difficulty in understanding how to manage conflict within a project. Expressions of conflicting points of view must be encouraged to nurture openness and honesty in any organization. Leas[9] argued that, although there are times to curb conflict, there are also times to instigate conflict for the good of the organization. He noted the following reasons conflict should be escalated rather than decreased.

1. People are so caught up in being nice and agreeable that they do not look at problems seriously or are not challenged by ideas.
2. People wanting harmony and peace make it difficult for anyone who is not like them to become part of the organization. Hence there is a tendency to promote conformity rather than an honest discussion of ideas.
3. When differences and uniqueness are accentuated, aggressive behavior is minimized. If people feel free to express themselves, they feel less disenfranchised, and therefore are better able to work with others toward a manageable solution.
4. In moderate amounts, conflict is a way of expressing aggression. It is better to have this aggression expressed openly than to hold it inside until there is a volcanic explosion.
5. Finally, conflict increases consciousness, aliveness, and excitement.

Although writing from a business perspective, Blackhard and Gibson[10] noted that opportunities emerge when leaders learn how to capitalize on conflict. They stated,

> Conflictive behavior in the workplace [or community] can range from very positive at one extreme to very counterproductive at the other. Properly managed, conflict can enhance creativity through constructive challenge and interchange, improve decisions by introducing more information and perspective, and foster learning through mutual problem solving. It can therefore further the purpose of the organization by improving the performance of its people and systems.

---

[9]Leas, S. B. 1982. *Leadership and Conflict*, pp. 107–109. Nashville, TN: Abingdon Press.
[10]Blackard, K. and J. W. Gibson. 2002. *Capitalizing on Conflict: Strategies and Practices of Turning Conflict to Synergy in Organizations*, p. ix. Palo Alto, CA: Davies-Black.

These points are important to understanding why managing conflict, in contrast to squelching or controlling it, is essential to complex community development projects.

Strategy for Success: Make Everyone Feel Welcome and Safe

Fear of regulation and trust in local leadership represent the combination of conflict and politics that gave the Maquoketa watershed citizens an incentive to attend watershed meetings and an opportunity for successful action. Residents attended the initial meetings for various reasons. Many had never participated in any community events. In one case a farmer came without his brother, his business partner, who refused to participate in this community meeting because the "government has no business telling us what to do." A couple of farmers came together united by a common concern, even though it was well known that they had personal disagreements with each other. Several farmers indicated that they would not participate beyond this meeting, but wanted to see firsthand what was taking place. Curiosity about the purpose of the meeting and fear that decisions would be made without their input drew in several producers. There was also a large group of farmers who were known to be very conscientious producers and who wanted to be part of any decision especially if it might impact conservation programs. Residents of the small towns in the sub-watershed also attended in significant numbers.

Pulling so many competing interests together increased the potential level of conflict. It was critical to make every resident feel welcome and safe for them to fully participate in the political process required in these watershed meetings. The best strategic decisions made for each of these sub-watersheds was identifying and inviting local leaders in the watershed to participate on the steering committee. These local leaders provided guidance on how to reach out to as many residents as possible, even those with combative personalities. Another strategic decision was to have the watershed councils develop their own meeting agendas, and develop leaders who could manage the agendas during the meetings. Keeping focused on the goals of the meeting was essential to the success of the meetings.

The types of conflict as described above show that conflict is not one-dimensional. Conflict has many different faces that can arise at unexpected times and in unanticipated ways. During community meetings, when an individual expressed frustration or anger, facilitators and project leaders were never sure if it was because of something going on during the meeting or elsewhere in their lives. Yet conflict is a form of capital that, when reinvested and placed in its proper perspective, results in a stronger project with a greater likelihood of success. Without the threat of regulation and the promise of additional conservation funding assistance, the citizens of the Maquoketa watershed might never have come together to initiate their project. As capital, conflict served as a source of energy that invigorated the community. Meeting attendance remained strong, and, for the first time, every issue and idea was argued in a public setting where everyone was welcomed. As a result, final decisions and strategies embodied everyone's input, even though some perspectives carried more weight than did others.

The issues that create conflict and tension in controversial projects, such as mobilizing farmers to take control of their watershed, also create conflict and tension for the community development specialist, whose job requires remaining neutral as an outside facilitator so conflict can emerge and then be dealt with openly and constructively. It is easy to "side with" key community leaders or government representatives when the specialist knows they will have to work with these key individuals on future projects. However, this can result in the community seeing the extension worker as a representative of government, rather than a fair and knowledgeable educator who can be trusted to serve the community first and foremost. While it is not easy, the challenge for the community developer as outside facilitator is to remain neutral and provide space so conflict can emerge and be dealt with openly and constructively. Using conflict as an asset to watershed council development requires delicate diplomacy, an understanding of the political dynamics, and knowing when to let the group confront its conflicted issues and maintain order.

Acknowledgment I would like to acknowledge the contributions of John Rodecap, Maquoketa Watershed Project Coordinator with Agronomy Extension, and Lois Wright Morton, Extension Sociologist, Iowa State University.

Chapter 6
The Language of Conservation

Jacqueline Comito and Matt Helmers

One April morning in 2008, we convened a listening session of a group of Natural Resource Conservation Service (NRCS) field staff from Soil and Water Conservation Districts in southwest Iowa. This group of conservation professionals seemed frustrated because they felt their message of conservation was ineffective compared with competing advice to increase yields and profitability. They were conflicted between wanting to defend and explain current practices and acknowledging that what is being done is not nearly enough to improve water and soil quality. The hour-long session ranged from concerns about increased erosion due to recent trends of increased tillage to complaints that urban residents did not appreciate the job farmers were doing to protect the waters in the state.

The frustration was palpable when one gentleman interjected, "I think we need to re-language some of our conservation practices so other people will understand them and kind of relate a little bit better."[1] Could "re-languaging"[2] be a key to increasing conservation practices and improving water and soil quality? If we change the language, will that encourage changes in individual practices? Can changes in language create as well as reflect value changes in a society?

This chapter describes the results of listening sessions with conservation and watershed field specialists and conservation-minded farmers. The purpose is to let their voices be heard and explore how these voices can change the discourse about

---

[1] NRCS technical specialist, 04040308.

[2] For the purposes of this paper, we will use the term "re-language," since it occurred organically during a listening session. The use of "language" or "re-language" is better identified as "discourse," the institutionalized way of thinking that is realized or made real through language. "Discourse" defines socially acceptable speech. Discourse is not limited to words but include all of the signs utilized by a society to communicate and direct our way of seeing issues and giving meaning to our actions and ourselves.

J. Comito (✉)
Project manager for the Iowa Learning Farms, Department of Sociology,
Iowa State University, 103 East Hall, Ames IA 50011-1070, USA
e-mail: jcomito@iastate.edu

L.W. Morton and S.S. Brown (eds.), *Pathways for Getting to Better Water Quality: The Citizen Effect*, DOI 10.1007/978-1-4419-7282-8_6,
© Springer Science+Business Media, LLC 2011

farming and the environment. We look at how farmers, Iowa State University Extension (ISUE) agriculture experts, Iowa Department of Natural Resources (IDNR) field staff, and NRCS field staff see the issues affecting conservation, their roles as conservationists, and their ability to influence change. The qualitative material for this analysis comes from 15 Iowa Learning Farms (ILF) listening sessions held with the above groups in 2008 and 2009. We frame these issues using discourse analysis and then discuss what we heard in the listening sessions to better understand the conservation messages that are being delivered by agricultural conservation stakeholders.

Does the "Language" We Use Matter?

Can we "re-language" our social interactions and get changes in behavior? Does the "language" we use matter? Scholars working in discourse analysis take a dynamic view of language and its meaning, not in terms of dictionary definitions but as something socially negotiated. Bakhtin[3] pointed out that "it is not, after all, out of a dictionary that the speaker gets his words" rather the speaker hears them "in other people's mouths, in other people's contexts, serving other people's intentions."

The words we use define who we are and what we do, bringing our identities into existence. Discourse[4] is a social occurrence in which every utterance is socially, politically, and historically contextualized.[5] So the past, or memories of past behaviors, is one of the key factors used to give meaning to any given moment or action. It is here that the tensions between the identity-defining functions of discourse and the content of discourse can be exposed and realized. For example, asking people to recall and talk to others about a personal experience with water (such as fishing, swimming, catching frogs and crayfish, or viewing blue herons wading in shallow streams) helps them to reattach meaning to the importance of protecting water. Discourse conveys meaning (content) and also allows individuals to define their own identity in the present context. Through discourse, individuals assume a "responsibility for inventing themselves and yet maintain their sense of authenticity and integrity."[6]

---

[3]Bakhtin, M. M. 1981. *The Dialogic Imagination: Four Essays*. Austin, TX: University of Texas Press.

[4]Discourse occurs whenever two or more people are gathered together around a given idea or social issue such as farming or agriculture.

[5]Bakhtin, M. M. 1981. *The Dialogic Imagination: Four Essays*. Austin, TX: University of Texas Press.

Voloshinov, V. N., Matejka, L., and Titunik I. R. 1986. *Marxism and the Philosophy of Language*. Cambridge MA: Harvard University Press.

[6]Myerhoff, B. G. 1992. "Life History among the Elderly: Performance, Visibility, and Remembering." P. 232 in *Remembered Lives: The Work of Ritual, Storytelling, and Growing Older* edited by M. Kaminsky. Ann Arbor, MI: University of Michigan Press.

We can gain a more complex understanding of power relations and other social tensions in a group by examining how a speaker, through contextualization techniques, implicitly signals to his or her listeners how the speaker's narrative should be interpreted.[7] The interpretation of any given discourse rests in its linkage with other memories or mythologies and in its contextualization.[8] When people evoke the past in a present moment, the assertion of themselves as social beings reflects a continuance to a world that is already in motion. This dynamic is true in politics; it is true in agriculture; it is true in any situation that involves two or more people. Mannheim and Tedlock[9] argue that "no one can speak or write language, as we now know it, without already being situated in this world." The same is true for the non-verbal signs used in any communication event – the participants often give indications of how the signs can and should be understood. The meaning of discourse rests in-between participants whose relationships to each other are often in a state of flux as the participants are simultaneously spectator and spectacle.[10]

The Good Farmer

The power of discourse is evident in the ways business and marketers use visual and verbal messages to move people to action. The agricultural industry uses nostalgic images of the "perfect" farm to sell their various products, such as seed and chemical applications. The farmers depicted are younger than age 40 years, with fields of healthy beans or corn growing out of finely plowed black soil. This image endures despite the fact that the average age of Midwestern farmers is late 50s and the uncovered, tilled soil is known to diminish water and soil quality through soil displacement. The industry uses those images because they tap into a socially acceptable discourse on farming practices.

---

[7]Basso, E. 1990. "Introduction: Discourse as an Integrating Concept in Anthropology and Folklore Research." Pp. 3-10 in *Native Latin American Cultures Through Their Discourse* edited by Ellen Basso. Bloomington, IN: Folklore Institute, Indiana University.

Goffman, E. 1974. *Frame Analysis: An Essay on the Organization of Experience.* Cambridge, MA: Harvard University Press.

Irvine, J. T. 1996. "Shadow Conversations: The Indeterminacy of Participant Roles." Pp. 131–159 in *Natural Histories of Discourse* edited by M. Silverstein and G. Urban. Chicago, IL: University of Chicago Press.

[8]Graham, L. R. 2000. "The One Who Created the Sea: Tellings, Meanings and Inter-Textuality in the Translation of Xavante Narrative." Pp. 252–271 in *Translating Native American Verbal Art: Ethnopoetics and Ethnography of Speaking* edited by K. Sammons and J. Sherzer. Washington DC: Smithsonian Institution Press.

[9]Mannheim, Bruce, and Dennis Tedlock. 1995. "Introduction." P. 7 in *The Dialogic Emergence of Culture* edited by D. Tedlock and B. Mannheim. Urbana, IL: University of Illinois Press.

[10]Drewal, Margaret Thumpson. 1992. *Yoruba Ritual: Performers, Play, Agency.* Bloomington, IN: Indiana University Press.

Conservationist farmers and natural resource technical professionals have to first negotiate and change the perception that black soil represents good farming if operators' tillage practices (behavior) are to be effectively changed. The challenge of changing the meaning of perceptions is illustrated by a southeast Iowa farmer who expresses his concerns about the appearance of a no-till field, with its high level of residue, compared with a tilled field:

> One of the biggest hurdles to me [in convincing others] is the way the field looks from the time the corn or beans emerge until they cover the ground. It looks like crap. We were told when I went to no-till conferences, the first thing you do when you're done planting is go fishing, just go away, just go away for a couple weeks and then come back later. Because you drive by a hill that's been turned black, it's got corn in it, the corn comes up faster, it's darker green – now, it doesn't mean it's going to yield more, but it does look better; I mean, there's no question about it.... But that is when a farmer brags about his field, when it's coming up. You don't brag about a field in October.[11]

In the end, the farmer told us, tall beans and corn do not make good beans or corn. This farmer's story negotiates and changes, if ever so slightly, the preexisting ideas, social relationships, values, and beliefs concerning "good" farming. The "black" field is part of the social discourse of farming and as such is socially, politically, and historically contextualized.

The Listening Sessions

The listening sessions emerged from discussions of our work at the ILF, a conservation initiative started in 2005. ILF is a partnership among conservation-minded farmers, the ISUE, the Leopold Center for Sustainable Agriculture, the Iowa Department of Agriculture and Land Stewardship (IDALS), IDNR, NRCS, the Iowa Farm Bureau (IFB), and the Conservation Districts of Iowa (CDI). The overall goal of the listening sessions was to strengthen understanding about individual farm-level decisions and the impact on the environment. We wanted to get a better handle on how the local stakeholders think about water quality concerns. Less formal listening sessions with farmer/local stakeholder groups in the five distinct soil regions in Iowa were conducted since the beginning of ILF in 2005. The purpose of those sessions was to understand the conservation needs and concerns of farmers and local stakeholders and adjust ILF programming to better meet these needs.

In 2008 and 2009, we teamed with Dr. Lois Wright Morton, ISU Sociology Department, and expanded the study to include staff from conservation districts in the five NRCS areas and IDNR field offices. Attendees at the 15 listening sessions were 38 farmers, 85 other community members,[12] 134 NRCS field staff, and 51 IDNR

---

[11] Farmer, Keokuk County listening session, keo020108.

[12] This group consisted of Iowa State University Extension agricultural professionals (such as field agronomists and program specialists), NRCS District Conservationists, watershed coordinators, teachers, county naturalists, and local Soil and Water Conservation District commissioners.

field staff. In 2008, the first listening session was held in January and the last session was held in July (after Iowa experienced extensive flooding throughout central and eastern Iowa). The IDNR listening sessions were held in January and February 2009. Each group was asked the same series of questions, which they had been given in advance of the session. A facilitator provided discussion prompts centered on conservation management systems and water quality themes. Prompts were specifically open-ended and flexible to allow the participants freedom to explore different avenues of information not anticipated by the researchers. All of the meetings were recorded and transcribed for analytical purposes.[13] One of the key themes of these sessions was the role language played in promoting or undermining conservation practices that protect water.

Is There a Water Quality Problem or Not?

Soil erosion is tangible – I can see it, I don't want it, I've got to drive over it, I want to do something about it – and the fix is tangible. Water quality is not tangible – I don't see it, feel it, taste it. The fix isn't tangible; it may be doing something, but I can't see it, and it doesn't affect me directly. People down in the Gulf that fish shrimp are not my concern – I raise swine.

NRCS field specialist,
02010908

...And sometimes you've got to take them down there and run the test. You've got to take an ammonia kit down there and say that the runoff from that feedlot is coming down here and impacting the water quality. Sometimes it takes that; otherwise, they just think it's the neighbor up the road. So getting them out in the field, showing them what's going on. Otherwise, they just think it's coming further upstream. And that helps. We've had to take guys out and have the field test kit and get their eyes opened up, like, all right, that's my problem. What do we have to do about it. A lot of them think that it's not just them....

IDNR field specialist,
05013009

Farming the Government

One of the main goals of the listening sessions was to gain a better understanding of how local stakeholders view water quality issues. This is a difficult question that, when answered by people from state agencies, educators, and farmers, reveals a number of unresolved internal and external conflicts. Today's agriculture measures its success by higher efficiency and higher yields, often failing to factor in

---

[13] This study was funded through the Iowa Learning Farms program. Heartland Regional Water Coordination Initiative paid for the transcription of the listening sessions.

environmental responses such as soil erosion, water pollution, and flooding – changes that can eventually collapse the overall system. Agriculture is a complex industry with a multitude of factors affecting everyday land management decisions. The amount of money in the industry invested in conservation pales when compared with the resources allocated to sell yields, chemicals, and equipment.

The whole industry is filled with mixed messages. The most obvious case is the current United States Department of Agriculture's (USDA) farm programs that promote higher yields and productivity through crop payments while simultaneously funding, and in many cases underfunding, conservation programs geared toward encouraging and rewarding good conservation practices. One of the stakeholders attending a listening session explains:

> I've done a little policy work and I've coined a term for lack of a better one, called "bureaucratic inertia." ... We've had for generation after generation in this country a policy of cheap food and it's been no secret about that ... we've often wondered if it wouldn't be more beneficial, both from a farming standpoint and a water quality standpoint, to diversify and give farmers an extra couple of streams of income using plants and materials that maybe aren't necessarily so till-intensive, chemical-intensive and things of that nature.... But a lot of that policy comes from Washington. I mean ... corn, soybeans, some cotton, some millet and wheat gets some attention. But for the vast majority, we're looking at, we're still propping that cheap calorie system up. And I think that really addresses some of your past concerns – it's the way Grandpa did it; it's the way whoever did it. But really in the framework that we're currently faced with, do we really have a choice? Can we branch out? No. Well, there's not a payment for that, whatever. And of course the economic standpoint – well, then I can't afford to make a living – so it's back to the corn and beans or the cotton or whatever.[14]

What this speaker is acknowledging is that while farmers will say they are following what their fathers did, their management choices are also shaped by the USDA farm policy. Or, as an IDALS employee recently said, "We have taught them how to farm the government. Farm policy has to change if we want to do anything about water quality." Current social definitions of "successful" farming are based on yields rather than profit, with most of the financial accounting failing to factor in the true costs of all inputs and outputs, including soil loss and decreased water quality.

The schism in USDA's policies between "feeding the world" and "protecting the land/soil" can create conflicting discourses and reinforce competitive rather than complementary goals. The dilemma and challenge are to create a vision and on-the-ground applications that integrate agricultural production with protection of natural resources by those agencies that are responsible for promoting conservation: IDALS, IDNR, Soil and Water Conservation District commissioners, universities, and local technicians.

---

[14]This was an area lawyer who was attending the meeting with his farmer father. Fay020808.

Lack of Urgency

Analyses of our 15 listening sessions portrayed ambiguity about how water quality issues are perceived in Iowa. The farmers who attended our sessions were considered by others in their areas to be strong conservation farmers. The other attendees were county and state employees with responsibility for improving water and soil quality in Iowa. The group discourse revealed concern about soil and water quality on the part of all who attended our meetings, but in many cases there was no sense of urgency. For both groups, farmers and agency specialists, part of the limits of acceptable speech in agriculture is to leave things ambivalent to avoid unnecessary or dramatic changes in production techniques due to natural resource concerns.[15] In other words, unless the evidence is overwhelmingly in favor of change, status quo is reasonable and preferred.

When asked whether there was a water quality problem in their area, not even the NRCS technical specialists were prepared to say that there was. Most participants did agree that progress had been made over the last several decades except in the last few years. This response by a Conservation District specialist is typical of the responses we heard:

> It's a real subjective thing to ask because most all of us don't have any benchmarks. We don't know. There's no monitoring. And what aspect of water quality? Are you looking at pharmaceuticals, are you looking at bacteria, are you looking at nutrients, are you looking at sediment? What aspect? Water quality is a broad brush. And here we've got a watershed that we've got quite a bit of monitoring in; that's only in the last five years. We have a recent benchmark, but the science isn't there to really be able to give you a very good answer to that question.[16]

While our participants seemed to believe that the lack of long-term data negated the urgency of current water quality issues, it does not seem to prevent them from asserting that progress has been made. One of the possible explanations is that part of what is socially acceptable when discussing farming practices with agriculture stakeholders is to make certain one acknowledges the substantial progress made in reducing soil erosion during the last 30 years. The message of this discourse suggests that status quo is acceptable and "we are doing okay because we haven't had another dust bowl." Would a conversation that acknowledges the concerns of environmentalists (often expressed in negative terms in our listening sessions), while appreciating the progress made by farmers, better prompt conservation goals and lead to more explicit discourse (and action) about the need for a much stronger effort if water quality is to be achieved?

The last NRCS listening session was held in July in southeast Iowa, one of the areas most affected by the Upper Mississippi River basin spring and summer flooding of 2008.

---

[15] It can be argued that this ambiguity does not rest with the part of the industry that promotes new products such as higher yielding seed, new chemical applications, or the latest in equipment.
[16] NRCS technical specialist, 0210908.

It was clear that the widespread flooding brought about a greater sense of urgency in the minds of the NRCS field staff. Perhaps for the same reason, during the January and February 2009 sessions conducted with IDNR, that field staff was also a lot more assertive of the water quality concerns in Iowa and equally assertive that it was going to take strong regulations and societal investment to make it happen:

> I would like to see all watersheds in Iowa having to meet strict water quality standards for silt runoff, nutrient pollutants, runoff water, volume, etc. If the standards are not met, mandatory improvements will be required for the landscape, such as wetland restoration from the … vegetative cover. All tile water needs to be captured and held for retention, flood prevention and nutrient pollutant removal. We will need government programs in place to pay for these mandatory permanent conservation easements so the cost is not on the landowner. Farming and livestock production is industry and should be regulated as such.[17]

The conversations during that session had more clarity and were forceful in advocating the value of conservation measures in contrast to the messages we heard in the nine sessions held prior to the late spring flooding. Without extreme events, most stakeholders hold on to what is tangible and what can be seen – soil erosion.

Soil erosion has been the "public" message to farmers since the 1930s. Farm survey respondents typically rate soil erosion as a top priority even if they do not actually practice preventative measures. While conservation planning is based on a "tolerable" level of soil loss (T), farmers often receive mixed messages about this standard. At a field day in 2007, an older farmer asked the NRCS representative if managing for "T" was still good enough. The NRCS's answer was "yes" and "no" and he spoke of moving beyond "T" and building on soil organic matter or soil carbon. The farmer seemed unclear as to what to do with the answer he received.

We came to several conclusions after reviewing all of the listening sessions: (1) there is still a need for greater understanding on the part of both agency staff and farmers of the impacts of land management on water and soil quality; (2) more demonstration and other participatory learning opportunities are needed to further the understanding of conservation;[18] (3) messages need to be inspirational enough to increase stakeholders discourse and move them to action; (4) all three groups felt like they were not being heard by the other groups and by the "folks in Des Moines," leaders and bureau chiefs for IDNR, IDALS, NRCS, and state legislators; and (5) the aftermath of the 2008 floods emphasized the urgency of increasing conservation practices to protect land and water resources. The first step to "re-language" conservation is to recognize the internal and external conflicts and confusion caused by the complexity of the problem. This complexity and confusion can discourage discourse

---

[17] IDNR field specialist, 03021609.

[18] Aldo Leopold in "The Land Ethic," a chapter of *A Sand County Almanac*, writes: "Conservation is a state of harmony between men and land." Leopold felt it was generally agreed that more conservation education was needed; however, quantity and content were up for debate. Almost 60 years later, we would have to agree with him; Leopold, A. 1949. *A Sand County Almanac and Sketches Here and There.* New York: Oxford University Press; Leopold, L. B. (ed). 1993. *Round River; From the Journals of Aldo Leopold* (Pp. 156–157). New York: Oxford University Press.

and lead to inactivity, reducing capacities to engage in changing behaviors. Public discussions and policies that engage farmers and technical specialists simultaneously can link the science of what is known with local knowledge to increase appropriate management practices under risk and uncertain conditions.

Building a Culture of Conservation

> When one considers the prodigious achievements of the profit motive in wrecking land, one hesitates to reject it as a vehicle for restoring land. I incline to believe we have overestimated the scope of the profit motive. Is it profitable for the individual to build a beautiful home? To give his children a higher education? No, it is seldom profitable, yet we do both. These are, in fact, ethical and aesthetic premises which underlie the economic system. Once accepted, economic forces tend to align the smaller details of social organization into harmony with them.
>
> There is as yet no social stigma in the possession of a gullied farm, a wrecked forest, or a polluted stream, provided the dividends suffice to send the youngsters to college. Whatever ails the land, the government will fix it.
>
> Aldo Leopold, *Round River*

Aldo Leopold believed that, in explaining why farmers do not practice more conservation, we have "overestimated the scope of the profit motive." After listening to the hundreds of local stakeholders explain the economics of agriculture and farmer decision-making processes, we have to agree. When we asked extension personnel, agency field staff, and local stakeholders "What were the three top factors affecting land management choices?" the answer was economics. When we pushed the question a little farther, we got these answers:

> NRCS 1: Yields have a lot to do with it. Economics aside, some people just like to have bragging rights. They want to be that guy that's the top producer in the neighborhood.... When you go to the coffee shop, they don't talk about, 'Well, I had a net of so many dollars per acre.' It's bushels or whatever unit of measurement ... pounds of beef, bushels of corn, bushels of soybeans. It isn't about their efficiency. And that is pretty much a direct result of what agencies have talked to them about for years. Increase your yield by doing this.
>
> NRCS 2: Yeah, We don't say, 'Hey, you're going to save three tons of soil today.'[19]

These individuals seem to understand the role they played in promoting yields over conservation.

When the same question was asked of farmers, they seemed to agree that farmers go for high yields more than they do for profitability, as evidenced by this conversation between two farmers at a listening session:

> Farmer 1: You know, back I don't know how many years ago the Fayette District started an economic yield contest and went on to be the max contest that Successful Farming had. But anyway we found out that in most cases the top yield was not the most profitable yield, that

---

[19] NRCS technical specialists, 02010908.

it's usually the second or third yield down. And in one case it was the exact opposite; there were only four entries in this one class, and it was the exact opposite—the highest yield was the least profitable and the lowest was the most profitable. But in most cases it was one of the closer to the top yields, but it very rarely was the top yield…. I think there's too many people that don't look at the actual economics of it. They say they're looking at the economics, but they're basically looking at the yield…. You have to be talking about what you end up in your pocket.

Farmer 2: Yeah, but high yield gets you the biggest smile in the coffee shop.

Farmer 1: Yeah, but that's about all it does for you.[20]

In the end, farmers were arguing that issues such as peer pressure, ease of farming, tradition, and legacy played almost as important roles in influencing their decisions as did yields and profitability.

How Do People Change?

What are the traits of those farmers and technical specialists whose discourse leads to their own changed behaviors and influence over others? What are the conditions that create "re-language" opportunities that can alter public understanding and behaviors? In 2007, we asked our ILF farmer cooperators what motivated them to become good conservationists. The following personal experiences seem to represent ILF farmer cooperators as well as the conservationists who attended the listening sessions in 2008:

- They were raised with a strong conservation ethic.

  I don't know why I have strong feelings about conservation. I always hated going through fields and seeing erosion channels. Always hated it. Maybe because I know how long it will take to get it back. Maybe because I am young, 30 years old, and I worry what the land will be like when I am older. Once it is gone, it is gone. Even if no-till didn't save me time, I would do it. You always hear people say this generation does something on their farm because their grandfather or father did it. I have never done anything because that is how it has always been done. I am the one that drives the conservation on the farm. I have to do it on my operation first and if it works then my father will do it.

- A major event motivated them to do something different.

  My ethics were that no-till was the right thing to do and I was going to try and stick with it until it worked. There were years when there were disasters but after I made adjustments to planters and equipment, I was able to make it work. I first started no-till because in spring 1979, we had a heavy spring rain and there was bad erosion and I figured we needed to change that. A big rain event got me thinking we needed to make changes.

- Took a class or read a book that motivated them to do things differently.

  In 1978, I went to ISU that winter and took 19 college credits. One of the professors held up a book on ridge tillage. I bought that book and read it cover to cover. I came home and

---

[20] Two farmers, Fay020208.

told Dad I am going to do no-till. My dad thought I was crazy. But I started with 10 acres. I made some mistakes but it worked. When Dad saw that he started doing some acres. Our neighbors thought we were crazy. This farm has really changed since then. We had lots of erosion spots and waterways washed out but now you can drive over the whole farm without a problem. Now most of the neighbors are doing no-till.

- A change in regulation motivated them to do something different.

In 1988, the Farm Bill said we had to go with more no-till. I needed a new planter so I bought one that could do no-till. I didn't think it would work but was going to give it a try. By that time, technology had come along that made it work. Round-up helped. Biotechnology caught up. Ten years into no-till we noticed organic matter was increasing. No-till works. I could go to anyone's farm in this area and make no-till work. And I am not the best farmer. You just need to stick with it.

These farmers are curious and willing to take risks, creative in finding solutions, and caring about the land and their communities. Their personal observations and experiences reinforce their conservation ethic and they become champions, changing the discourse of conservation and in turn shifting social norms in support of managing simultaneously for both productivity and protection of the natural resource base.

Regulations alone will not sustain change. In many cases, farm managers will reverse their practices the moment the regulation is lifted unless during the time specific practices are required, the farmer learns that the new practice is also advantageous. True change is inspired. Technology will play a role in the transition to a resilient agricultural system. However, the transition to a culture of conservation embodied as change in agriculture's discourse about the environment is the more daunting challenge. ISU Extension and agricultural researchers, IDNR, IDALS, NRCS, farmers, and the land managers need to listen to each other. Change can begin to happen when we cultivate listening – to really listen and hear so that the person who's talking feels understood:

I think years ago when I was on a water quality project … and I had a landowner [and] we looked at every aspect of his farming and asked him what he needed. And he told me the things he needed in his farming practice, and we came up with changing some of the guidelines, even in the NRCS Tech Guide, to fulfill what farmers needed. He came back, and over the next five years of that project, if I needed any assistance from him as far as going out and helping landowners in other counties, he'd go out, he'd get on the circuit; he would talk to those other landowners. He'd go out and help them put in their solar-powered high-tensile wire along their streams. He'd fence a lot of streams. Farmers didn't believe in fencing the streams because of the problems with fencing and all that. But he proved to them…. And finding somebody like that that wasn't a believer to begin with … but it worked for him. And he went out and kind of preached it to other people, and it really, to me that's a good steward – somebody that initially probably wouldn't have tried these types of conservation practices but gave it a try, worked for him. The things that didn't work, he'd tell you, but of course he didn't go out and spew that. But that's what I think is a good conservationist, someone who believes in what they're doing and promotes it.[21]

---

[21]IDNR field specialist, 05022009.

Change starts with listening to all the stakeholders (such as farmers, extension staff, IDALS, IDNR, and NRCS)[22] and understanding that the world might be different than we believe it to be and beginning to build our solutions from that new collective perspective.

We agree with the NRCS employee who argued that we should "re-language" our conservation practices in a way that inspires others to follow. Language is a powerful tool and the words we use do matter. We would encourage conservation stakeholders to evaluate the messages they are delivering to see if they are unintentionally misleading or confusing. Currently there is a good deal of confusion as well as inner and external conflict concerning conservation and its relationship to agricultural productivity. This uncertainty leads to inaction or status quo. Until people can see and articulate clearly the problem of water and soil quality and potential long-term consequences to themselves and their watershed, they will not feel the need to change.

We also caution against jumping on the "green" language bandwagon. Parts of the "green" campaign are based on false claims or "re-languaging" products (such as automobiles) or ideas without operationalizing the messages.[23] Thus along with "re-languaging," institutional structures must not only increase their efforts to give consistent messages but also find innovative and effective ways to provide resources and expertise in support of practices that protect water quality. In 2007, when the ILF program "re-languaged" around the idea of a culture of conservation, we also restructured our program to better meet our goals and the needs of agriculture in Iowa.

A Culture of Conservation[24]

The Culture of Conservation approach to ILF is an example of how "re-languaging" the way we discuss conservation can increase a program's visibility and effectiveness. During the past 2 years, evaluations of ILF reveal increased name recognition and landowner changes toward increased conservation practices. While ILF cannot take sole credit for these changes, it is clear that the idea of Building a

---

[22] In particular, it seems that IDNR, IDALS, NRCS, and ISU Extension need to work harder at listening to each other. Limited research and program dollars often place these groups in competition vying for power, credit, and dollars, with each group thinking they have the "best" solution for the state.

[23] For instance, Toyota has a commercial where they claim their cars are "green" and show an image of a car made of leaves and branches gently decomposing into the earth, eliminating their ecological footprint. This message is false and dangerous, implying that buying a Toyota is all one needs to do to "save the planet."

[24] We were aided in this section on ILF from members of our communications team: Jerry DeWitt, Paul Lasley, Carol Brown, John Lundvall, Jamie Benning, Xiaobo Zhou, and Jean McGuire.

Culture of Conservation is more compelling, succinct, and encompassing than our original vision statement: *The Iowa Learning Farm promotes efficient agriculture production systems that result in agronomic, economic and environmental improvements through increased awareness and adoption of conservation systems and ethics.*

In addition, we made structural changes so that our functional activities support building a "community" of conservation and reinforce our conservation culture language. We added communications expertise to our team so that the communications expert, anthropologist, engineers, agronomists, and economists are exchanging ideas on a regular basis and discussing conservation from multiple perspectives. This discourse deliberately "spills over" as a communication and programmatic strategy to actively increase farmer participation.

Rather than an expert-led initiative, ILF utilizes local leaders/farmers to educate and encourage others for continuing change to improve water and soil quality in Iowa. The ILF went from "renting" our 28 farmer cooperators land for research to giving them an active role in their on-farm demonstrations. This change has increased their engagement as spokespeople in promoting innovative ways to help all Iowa citizens have an active role in protecting and enhancing our state's natural resources. We are improving our outreach materials according to feedback obtained through continual dialog and review by participating farmers. Using the idea that "seeing is believing," ILF farmers are demonstrating cropping techniques that improve the water and soil quality on their land while remaining profitable. In addition to the farmer cooperators, ILF involves a broad set of stakeholders, partner organizations, and agencies. Our vision is much shorter now: *Building a Culture of Conservation ~ Farmer to Farmer: Iowan to Iowan.*

A Culture of Conservation involves strengthening our commitment to a set of values, beliefs, and attitudes about the importance of natural resources to our standard of living and quality of life. It recognizes good conservation practices, encourages others to begin conservation farming, and involves everyone. Overall, we believe the strength of ILF is that we integrate agronomic, economic, environmental, social, and cultural information from Iowa State University and other research institutions with the peer-to-peer messages. Our ILF farmer cooperator, Randy Caviness, a 20-year "never-tiller" said it best,

> Many people talk about building up their soil, but then they don't do anything about it. No-till actually does something about it. The more you leave the soil alone, the better it does. Some people think that if they till every four or five years, that they will get benefits of no-till. No-till should be called never-till. Economically, you don't give up anything to do no-till. We want to show that you can be profitable and you can save the soil. Quality soil has got to be the bottom line. People need to think about their soil and do a better job for future generations. You can't just think or say that you are doing a better job, you need to actually do a better job.

That message is clear and Randy would be happy to come to your land and show you how it can be done. He told us if he can do it, anyone can do it.

Appendix: Listening Session Prompts

The Iowa Learning Farms (ILF) Program, in conjunction with Iowa State University Extension (ISUE), focuses on conservation management systems and water quality. To develop education and outreach to improve our soil and water quality, we need to get a better understanding of current practices and the real issues facing conservation. You will help us by sharing your experiences and opinions about conservation practices in your area. On [date], we will be at your area meeting to discuss these questions. Your participation is voluntary and the information you provide will be summarized by soil region so that your individual opinions are not disclosed.

1. What tillage and related soil conservation management changes have you seen on the land during (given period?): Positive? Negative?
   Do you think, in general, we are moving toward increased or decreased implementation of soil conservation practices?
2. Why do you think these changes have occurred?
3. What are the top three factors you think farmers base their land management decisions upon?
4. If you had a "conservation wish list," what changes or enhancements to existing policy would you like to see implemented in your district, area, across Iowa, and why?
5. Can we prioritize those wishes? Who do we need to engage and how? How should the ILF team move forward in our campaign to build a "Culture of Conservation?"
6. Is there an effective way to "target" buffer strip and/or wetland reserve cost-share payments or land set-aside payments to maximize water quality enhancement?
7. What do you think are the characteristics of landowners and watershed residents' conservation ethic that would get us to better water quality and soil protection? That is, what does a preferred conservation ethic look like and what are the subsequent actions that occur because of it? Give an example of a farmer in your district that you think has a conservation ethic you'd like to see replicated. What does he/she do?
8. What kinds of things do you do in your work to foster a conservation ethic in landowners/farm managers you work with? What kinds of tools or strategies would help you do this more effectively?

Part II
The Data

Chapter 7
Measuring the Citizen Effect: What Does Good Citizen Involvement Look Like?

Linda Stalker Prokopy and Kristin Floress

Although citizen participation is commonly recognized as important to the long-term success of nonpoint source (NPS) mitigation projects,[1] public agencies and organizations generally lack adequate ways to define types of participation and indicators to measure the success of the participatory process. This gap in knowledge can lead to inadequate planning, management, and evaluation of the public participation aspects of watershed project plans. In this chapter, we describe a way to measure the quality of citizen participation in watershed efforts. This process can help agency staff design a project that has a desired level of citizen involvement and to track and evaluate citizen involvement over time in a watershed project.

Without citizen participation, projects can fail to meet their environmental goals because they fail to meet people's needs, their programs fail to attract participants, or they initiate change that is not sustainable and fall apart when resources are unavailable. The motivation of project leaders and watershed coordinators for increasing citizen participation can be primarily utilitarian or egalitarian. A utilitarian view of participation is that it is good because it will lead to better technical and adoption outcomes and ultimately better water quality. An egalitarian view is that participation is good regardless of short-term water quality outcomes if people learn, social capital is increased, and people are empowered. In the middle of these two perspectives is the notion that the process needs to be positive for participants if participation is going to lead to sustainable water quality outcomes. With this lens, participation needs to both accomplish short-term goals and foster human capacity. To that end, it is important to measure the quality of a participatory process. Measures of citizen participation are valuable for progress reporting to funding agencies especially if the amount and quality of participation are explicitly included

---

[1] United States Environmental Protection Agency (USEPA). 2008. *Handbook for Developing Watershed Plans to Restore and Protect Our Waters*. USEPA. Office of Water Nonpoint Source Control Branch. EPA 841-B-050005.

L.S. Prokopy (✉)
Department of Forestry and Natural Resources, Purdue University, 195 Marsteller Street, West Lafayette, IN 47907, USA
e-mail: lprokopy@purdue.edu

L.W. Morton and S.S. Brown (eds.), *Pathways for Getting to Better Water Quality: The Citizen Effect*, DOI 10.1007/978-1-4419-7282-8_7,
© Springer Science+Business Media, LLC 2011

in project objectives. Project efforts often generate empowerment/capacity outcomes before physical implementation outcomes have occurred.

Frequently federal and state funded projects require citizen participation for agencies and nonprofit organizations engaged in NPS mitigation efforts. Engaging citizens can be a daunting task for many agencies and organizations, including Cooperative Extension. Many watershed specialists are not trained in the social sciences and they do not know how to engage the public. More critically, agency staff are not trained to be able to recognize unsuccessful participation when it happens and so fail to adapt and improve. Under these circumstances, meeting funding requirements for public participation has the possibility of becoming "busy work" that is done simply to report that it has been done. The indicators of participation that are reported or measured in these cases can include: number of public meetings held, number of workshops conducted, and number of volunteer monitoring sites maintained. None of these indicators address the quality of participation.

We propose a practical way to measure the quality of citizen participation in watershed efforts that can be used by field staff whether or not they have been trained in social sciences. The process we propose covers the depth and breadth of participation. Many scholars have developed hierarchies of participation. These are frequently illustrated as ladders where higher rungs indicate increasing levels of partnership between the agency and citizens. For example, in Arnstein's classic ladder of participation, the lowest levels are classified as non participation, middle levels as tokenism, and upper levels as degrees of citizen power.[2] Implicit in this ladder is the assumption that citizen power is the highest and therefore best form of participation. Also implicit is that consulting with the public and informing the public are forms of tokenism. These are only tokenism if the intent on the part of the agency is to ignore citizen input.

In thinking about NPS projects in the United States, we find it makes more sense to think about different forms of participation without an implied hierarchy. The right level of participation for a particular project will change over time and as circumstances change. In this chapter, we explore the issue of participation through the lens of an agency that is trying to improve water quality. In many cases, agencies establish watershed groups or steering committees to help either inform or co-manage the watershed effort.

Stakeholders

Before we present different types of participation, we need to think about who needs to participate in an NPS project. Engaging the right stakeholders is clearly an important component of successful participation. A recent meta-analysis found that broad representation that involves all stakeholders is a key to the success of

---

[2]Arnstein, S. R. 1969. A Ladder of Citizen Participation. *Journal of the American Institute of Planners* 35(4):216–224.

collaborative resource management.[3] Stakeholders include anyone who will be directly or indirectly impacted by the NPS project. Stakeholders also include people who can impede project success. The USEPA's Handbook for Developing Watershed Plans to Restore and Protect Our Waters (pp. 3–5) suggests five categories of stakeholders for developing a plan, which we modify slightly below to make them applicable for all groups regardless of whether they are writing a plan:

1. People who "will be responsible for implementing" any actions
2. People who "will be affected by implementation" of any actions
3. People who "can provide information on the issues and concerns in the watershed"
4. People who "have knowledge of existing programs or plan"
5. People who "can provide technical and financial assistance"

The exact nature of the types of community members that will fall into these categories depends upon the particular watershed. However, likely people include farmers, homeowners, renters, recreationists, municipal officials, school teachers, business owners, and developers. In many watersheds, it will be important to further break down some of these categories. For example, in a large agricultural watershed, it might be important to engage both large- and small-scale farmers and row crop and livestock farmers. It cannot be assumed that one farmer will represent all other farmers; especially different types of farmers. The same is true of other types of stakeholders.

The ideal participant will be someone who is respected by their community and communicates regularly with a range of people in the community so as best to represent others. A small number of such well connected and influential participants can consult with project planners and help identify other appropriate categories of stakeholders and individuals for their communities. A broad coverage by as many of the "right" people as possible will improve a project's chance of success.

Types of Participation

Below we present six different types of participation that watershed projects may typically encounter: cooptation, nominal participation, program participation, participation in predetermined activities, consultative participation, and participation in decision making. For each of these participation types, a project may have a number of different types of activities. Some funding agencies require reporting the number of activities but, for the purposes of understanding community involvement, it is much more important to understand *who* participates *in what ways* rather than how many opportunities they have to participate.

The following types of participation are not mutually exclusive and groups can have people participating at all or some of these levels at the same time. The same people can participate at multiple levels simultaneously. The important idea is to understand

---

[3]Moote, A., and Lowe, K. 2008. *What to Expect from Collaboration in Natural Resource Management: A Research Synthesis for Practitioners*. Flagstaff: Ecological Restoration Institute.

that there are different levels that contribute in different ways to achieving project goals, and to make sure you have a diversity of people participating at the levels at which you desire participation. We illustrate the types of participation with examples from our own research and observations in Indiana and around the Midwest.

Cooptation

The form of participation called cooptation is really a type of non-participation or tokenism in which citizens are invited to participate only so that project organizers can inform them about decisions that have already been made. An example of this would be a watershed project developed at the state level and then presented to local audiences. Citizens are invited to a public meeting where the watershed problem is described and the predeveloped action plan is presented. There may be an opportunity for questions and answers, but no real dialog because the sponsors do not expect to change their plan. If and when citizens become engaged with the watershed issues, they will feel disenfranchised and as if they were forced to agree. Cooptation (offering public involvement only to deliver information) defeats the purpose of participation and in the long run will negate the benefits that can accrue from citizen involvement. Based on our observation of watershed efforts in the Midwestern United States, we do not think cooptation is the norm. However, it is always something for which to be alert and to avoid.

Nominal Participation

Nominal participation occurs when people attend meetings but do not actively engage with decision making. This could take the form of not speaking up in meetings or not voting in decisions. Nominal participation is subtly different from cooptation as described above. In cooptation, the agency or other project sponsor is intentionally or unintentionally not giving participants a genuine voice. In nominal participation, participants could have a genuine voice and could influence project outcomes but they choose not to. If groups find that more than about 10% of regular attendees are involved in nominal participation, we recommend that additional steps be taken to engage and empower participants.

In the Eagle Creek Watershed in Indiana, participant observation revealed that some individuals attending the watershed group meetings did not take part in discussions and decision making. While they attended most meetings, it was difficult to determine whether they agreed with the decisions being made, and the group potentially lost valuable information through the nominal participation of some of its members. In the nearby Clifty Creek Watershed, observation of group meetings and interviews with participants showed that group members often did not take part in formal voting regarding group activities. Most often people refrained from voting for a particular activity if they disagreed with it, and one participant stated that, "No one ever votes no." While these examples could be construed as cooptation, neither of these cases necessarily reflects poorly on meeting organizers, who could have the best of intentions and desire active participation by group members, but are simply having a difficult time moving beyond nominal participation.

Efforts should be extended by project staff experiencing the same phenomenon in other watershed groups to draw these individuals out through the use of facilitation techniques designed for this purpose.[4] Meeting organizers can also help participants move beyond nominal participation by making sure that topics at meetings are appropriate to the audience. Organizers should also make sure the audience is being brought in to the process early enough to develop buy-in.

Program Participation

Program participation involves people participating in programs organized by others. Programs can include field days, cost share for practices, and attending an educational event among others. Loosely, this type of participation can be thought of as "what's in it for me?" participation, where the goal on the part of the participant is to advance themselves – through money, increased learning, and other benefits – and not necessarily because they share a watershed perspective. The diversity of stakeholders may be more limited for this type of participation than the other types. For example, a watershed project in an agricultural area may only be targeting programs at farmers. In other cases, a diversity of participants may be desired across a variety of programming types. For example, in the Eagle Creek watershed in Indiana, school children are targeted for age-appropriate educational activities, row crop and livestock farmers and horse operations are targeted for cost share programs and related educational events, and developers and municipal officials are targeted for development-specific educational events.

Participation in Pre-determined Activities

Participation in pre-determined activities includes water quality monitoring, helping with youth education, and "adopting" stretches of streams and rivers. Participants in these activities are not necessarily involved in decision making but show up to help for the "greater good." Many states have river monitoring programs in place, whereby citizens are trained to collect water quality information and submit it to a state database. The citizen monitoring program in Indiana, Hoosier Riverwatch, is sometimes used by watershed groups as a means of collecting low-cost monitoring data and as an educational tool. Groups usually take advantage of the efforts of people in the greater community rather than relying solely on the watershed group members for these types of activities. In the Clifty

---

[4] Kaner, S. 2007. *Facilitator's Guide to Participatory Decision-Making.* Second Edition. San Francisco: Jossey Bass.

Creek Watershed, group members are almost exclusively responsible for water quality data collection, limiting the possibility of using this activity as a means to broaden participation.

While adult participants may not join volunteer monitoring programs with the intention of taking a decision-making role, involvement in monitoring is actually a learning experience that can result in a greater connection to the natural environment. Monitoring and discussion of results reaffirm a watershed point of view and promote group ownership of environmental issues. As participants increase their local knowledge, they begin to have confidence in their right to participate in decision making. Citizen watershed groups that are given a meaningful role often want to conduct their own monitoring to confirm for themselves assessments passed down from agencies and researchers. This participation often prepares them to more actively contribute to decision making. Project staff can help set up partnerships with existing citizen monitoring programs to fulfill this need.

Another common activity that watershed groups may plan is an educational event for K-8 or K-12 students. These events require the participation of trained environmental educators. In the Eagle Creek and South Fork Wildcat–Kilmore Creek watersheds in Indiana, volunteer educators came from local high schools, middle schools, and through Project Wet contacts. Citizen participation in these events – both volunteer educators and attendees – helps to raise awareness of water quality issues in the watershed, and also encourages the formation of a sense of place among residents, ultimately assisting the watershed group in achieving objectives related to social outcomes. At the group level, however, there is little to no participation by volunteers in decisions to hold the activities or in design of the activities themselves.

Consultative Participation

Consultative participation involves asking people what they want with the understanding that the project sponsor is still the ultimate decision maker. There are a variety of ways to solicit ideas, including public meetings, surveys, focus groups, and websites.

For example, in the Eagle Creek watershed, the group regularly holds quarterly progress meetings that are open to the public and designed to reach out to residents who otherwise may not be involved in group activities. These meetings function as informational tools for residents inclined to learn more about their watershed, with time to give feedback to project managers about their concerns or opinions about the goals and activities of the group. Individuals who are not already at least minimally aware of the concept of watershed management or have concerns about water quality do not attend these meetings. The Eagle Creek method allows for continuous updating of interested watershed residents about the group's activities. Other watersheds may begin their planning processes with large-scale consultations, either in the form of surveys or meetings.

Surveys of watershed residents can also be conducted as a part of consultative participation. In several watersheds in Indiana, we have surveyed residents to measure their awareness of, and attitudes and behavior toward, water resources. This information

River Vision – a visioning session designed to elicit the values of watershed residents – was held in the Wabash River watershed as the initial step in a planning process. More than 100 people attended this activity and participated in small group activities that yielded valuable ideas to incorporate into the watershed plan to be developed. The advantage of large-scale meetings such as River Vision is the number of people providing input in the initial stage of planning. The Eagle Creek quarterly meetings, on the other hand, draw few participants but are an ongoing activity.

is being used to plan and refine watershed activities. The downside of surveys is the lack of a two-way communication, thus surveys are less likely to serve as an educational tool as well as a consultative tool.

Participation in Decision Making

Participation in decision making is also called co-management or collaboration, in which stakeholders are as involved in decision making as agency representatives. Much of the literature about watershed management and watershed groups touts this level of participation as the ultimate goal of public participation, although significant barriers to co-management may be present in any given watershed. Many agencies need to change their fundamental operating structure to engage the public in this way. There is some debate in the literature as to whether co-management and collaboration are the same thing, however, they are both forms of participation in decision making.

Participation in decision-making can follow either a self-governance or an agency governance model. In an agency governance model, the agency is frequently the driver of activities. Even in situations where the agenda is set in conjunction with a broader stakeholder base, the agency is still a full partner. In the Clifty Creek watershed, the group of citizen participants was given much of the power for decision making throughout planning and management activities. However, the agency rules were a significant barrier to the group's decisions actually being carried out. An agency that receives funding for watershed management is still beholden to regulations of their agency. For example, the agency may approve or disapprove incentives and cost share for particular practices as acceptable best management practices on a watershed resident's land, and a watershed group does not have the power to override the agency rules. In a self-governance model, the agency steps back significantly from agenda setting and decision making. In the case of Clifty Creek, the group was well suited to take on a major decision-making role during the initial planning stages when problems were identified and objectives formed. In this stage, the agency acted more as a facilitator of the process rather than an overseer. In later stages, when practices needed to be installed on the ground, however, the group was much less empowered to make binding decisions. Effectively, the group in the beginning stages was a good example of self-governance, and in later stages was agency governed.

Assessing Breadth and Depth of Participation

Below we present an approach that can be used as a tool by agencies for measuring the quality of citizen participation. Our approach relies on assessing the breadth of participation (diversity of stakeholder groups) and the depth of participation (the type of participation in which the stakeholders engaged). A number of other measures of the quality of participation have been developed in the literature but these are unnecessarily complex for the average watershed project.[5]

We suggest that agencies or organizations that are interested in assessing the quality of citizen involvement in their watershed efforts take the following steps. Groups should develop a table like Table 7.1 to assist them in working through these steps. While these tables may appear complicated, as goals get more refined and types of stakeholders are broken down, nonspecialists can easily collect the needed information. It is critical to do this with as much detail as possible to thoroughly evaluate and improve citizen participation.

1. Develop a comprehensive list of stakeholders and classify each type within each of the five stakeholder categories described above. As the organization convening the process, you will likely fall into one of the latter categories of stakeholders and you should include yourself as a stakeholder.
2. Articulate participation goals. Do you want citizens to have decision-making power? Or would you rather have people involved only in program, activity, and consultative ways? What percentage of the different stakeholders do you want involved in some way? These goals should become part of the project's objectives so that achievements can be reported.
3. Keep track of achievement of these goals. What percentage of different stakeholder groups are involved in what types of participation? This can be done primarily by observation and reviewing project records. To differentiate between nominal, consultative, and decision-making participation, it may be helpful to periodically interview regular meeting participants and ask them how they perceive their participation.
4. If participation is not where you want it to be, reflect, talk, and listen to stakeholders, and, if necessary, consult resources (community development specialists, state agencies, USEPA handbooks, facilitation guides) to understand how to improve the depth and breadth of particiation. To move beyond nominal participation, watershed managers should consult group facilitation guides or attend facilitation trainings. Learning how to draw out participants can be a very valuable tool and aid significantly in management efforts.

You may also need to revisit the feasibility of having the type of participation you had hoped to have. For example, in the Rayse Creek watershed in south-central Illinois, agency representatives assisted a watershed group in their initial stages of formation and then gave responsibility for decision making fully to the group. Unfortunately, the group required more assistance than was provided or available, and they struggled with accomplishing, or even fully forming, objectives. In this

[5]McCool, S. F., and Guthrie, K. 2001. "Mapping the Dimensions of Successful Public Participation in Messy Natural Resources Management Situations." *Society and Natural Resources* 14:309–323.

Table 7.1 Illustration of evaluating participation goals

Stakeholder	Type of stakeholder[a]	Desired type of participation	Goal	Are they involved?	Goal met?
SWCD	a, c, d, e	Activity	Participate in everything	Yes	Yes
		Consultative	Involved in all decisions	Yes	Yes
Health department	a, c, d, e	Activity	Help with field days	No	No
		Consultative	Attend committee meetings	Yes, but only nominal participation	No
Livestock farmers	b, c	Program	100% participate in cost share and/or field days	No	No
		Consultative	60% respond to survey	Yes – 15 of 50	No
Row crop farmers	b, c	Program	100% participate in cost share and/or field days	Yes – 20 of 150 at field day; 2 signed up for cost share	No
		Consultative	60% respond to survey	Yes – 35 of 150	No
Farm Bureau	c, d, e	Activity	Help with field days	Yes	Yes
		Consultative	Attend committee meetings	No	No
NRCS	c, d, e	Activity	Help with field days	No	No
		Consultative	Attend committee meetings	No	No
Extension	c, d, e	Activity	Help with field days	No	No
		Consultative	Attend committee meetings	No	No
Elected officials	a, c	Consultative	Attend committee meetings	No	No
General citizens	d, e	Consultative	Attend public meetings	Yes	Yes

[a]Responsible for implementing plan; b affected by plan implementation; c can provide information on issues and concerns in watershed; d knowledge of existing programs or plans; e can provide technical or financial assistance

case, co-management was not a feasible option and the agency was not able to provide the group with the staff and resources that would have been necessary for the effort to succeed. While a management plan was created in Rayse Creek, it is a descriptive plan with little prescription for pollution remediation.

5. Just like any adaptive management approach, revisit this process every few months to ensure that participation is on track.

Citizens may contribute to watershed improvement projects in many roles as program, consultative, and decision-making participants. Agency project managers can speed the progress and increase both short-term and long-term measures of success by improving methods of identifying and evaluating participatory processes. While stakeholder participation in watershed management can seem like an ideal but unattainable goal, the methods presented in this chapter will provide agencies an easily implementable method for both planning and evaluating participatory processes.

Evaluation Process Example

We illustrate these steps with a hypothetical example. Group X is a run by a Soil and Water Conservation District and is focused on improving the quality of water runoff from agricultural lands.

1. Develop a comprehensive list of stakeholders including the agency or organization that you work for.

    Group X came up with the following list of stakeholders: The Soil and Water Conservation District, the County Health Department, livestock farmers, row crop farmers, hobby farmers, local Farm Bureau, Natural Resources Conservation Service county personnel, county extension, county elected officials, and general watershed residents.

2. Articulate participation goals.

    Group X will be responsible for making the majority of decisions but they want input from different stakeholders along the way.

    They want all farmers in the critical area to be involved in program participation. There are 200 farmers in the critical area (including 50 livestock and 150 row crop farmers).

    They want NRCS, SWCD, the Farm Bureau and County Extension to be involved in activity participation. They see a role for each of these groups in planning and convening educational events.

    They would like all the stakeholder groups to be involved in consultative participation. They plan to have the other agencies participate in a consultative way through attendance at committee meetings.

    They do not envision any stakeholder groups (other than themselves) being involved in decision making.

3. Keep track of achievement of these goals.

    They evaluate participation over a 6-month time period. During this time, they conduct various activities including three field days and two public meetings. They hold ongoing committee meetings to plan for general activities. They also conduct a survey of all of the farmers in the critical area. The SWCD staff are involved throughout this process making decisions about what needs to be done. Below are examples of how the quality of participation in those activities was assessed.

(continued)

Evaluation Process Example   (continued)

Observing participation:

Field days: Based on gathering information from the participants, they know that attendees included at least 20 farmers in the critical area of the watershed. They were all row crop farmers. They tried to get diversity in participation in planning and convening the field days but only the Farm Bureau helped them.

Cost share: Two row crop farmers signed up for cost share.

Public meetings: Thirty people attended the meetings. These included a few farmers (all row crop) and general citizens. The public meetings engaged all participants and everyone offered ideas through facilitated discussions about what they thought should be done.

Committee meetings: They held three committee meetings during this time period. While someone from the Health Department attended one meeting, this person did not contribute anything. This is classified as nominal participation not consultative participation.

Survey: 50 of the farmers (15 livestock, 35 row crop) returned the survey. The survey asked questions about what types of cost share programs would be helpful to the farmers.

Classifying participation:

By compiling this information into a simple table (see Table 7.1), they notice that they have a number of gaps. Only two of the five groups they hoped to have involved in activity participation are actually involved. This means they have some weaknesses in terms of breadth of participation. A number of stakeholders are also not engaged in a consultative capacity. Stakeholders that have participated in a consultative way have done so at low levels, e.g., only 23% of row crop farmers completed the survey. They wanted two stakeholder groups to participate in programs, but only one (row crop farmers) is participating.

4. Reflect and consult resources to understand how to improve depth and breadth of participation.

Group X can clearly see that participation is not where they want it to be. They had a low response rate on their survey, people are not participating in committee meetings, and low percentages of farmers are participating in programs. They note that they have a real issue with livestock farmers not participating in programs. They decide to talk to the other agencies to ask why they are not participating in committee meetings and to learn how this can be improved. They also talk to some influential livestock farmers in the county to understand why they are gaining no traction with this group.

5. Revisit this process every few months.

Group X maintains a table like Table 7.1 as a living document and updates it every few months and revisits steps 1–4.

Chapter 8
Regional Water Quality Concern and Environmental Attitudes

Zhihua Hu and Lois Wright Morton

Water Quality as a Social Issue

Water plays a vital role in the functioning of the Earth's ecosystems. Polluted water has a serious impact on all living creatures, including humankind. It can negatively affect every possible aspect of human life: drinking, daily household needs, agricultural production, recreation, transportation, and manufacturing. Water quality problems, like all other environmental issues, are social problems. Attitudes and beliefs about the environment and water influence how water resources are used and underlie the social willingness to respond to water pollution. Efforts to address water quality issues can be better directed when interventions take into account how people think about the environment and frame water concerns.

Results from stratified random sample surveys of water issues conducted in 36 US states from 2002 to 2008 provide a snapshot of public beliefs and perceptions of water issues. First, general environmental attitudes and beliefs, their historical roots and associations with demographic and community characteristics are presented. Then, state and regional variations on environmental attitudes, perceptions of ground and surface water quality, willingness to learn about water issues, and behavioral changes are examined. Lastly, we discuss our findings and implications for regional and national public policies and community educational interventions that protect and conserve water resources.

Environmental Beliefs and Attitudes

In psychological terms, attitude generally represents "a summary evaluation of a psychological object captured in such attribute dimensions as good–bad, harmful–beneficial, pleasant–unpleasant, and likable–dislikable."[1] According to the

---

[1] Ajzen, Icek. 2001. "Nature and Operation of Attitudes." *Annual Review of Psychology* 52:p. 28.

Z. Hu (✉)
Department of Sociology, Iowa State University, 403B East Hall, Ames IA 50011-1070, USA
e-mail: zhihuahu@iastate.edu

L.W. Morton and S.S. Brown (eds.), *Pathways for Getting to Better Water Quality: The Citizen Effect*, DOI 10.1007/978-1-4419-7282-8_8,
© Springer Science+Business Media, LLC 2011

expectancy-value model, whenever we form beliefs about an object, we are actually evaluating the object. A person might have multiple beliefs and evaluations associated with one object, but the person's overall attitude toward an object is determined by the subjective evaluations in interaction with the strength of the association.[2]

Environmental attitudes are a collection of evaluative judgments about the use, function, and value of the environment in general and/or specific aspects of the environment. These attitudes range from the view that nature is a resource with specific human-centered uses and functions to the view that nature is at risk and needs human protection from uses and abuses.[3] Environmental attitudes guide individual citizens' and society-at-large interactions with nature. On an individual level, personal norms and beliefs influence how people approach the natural environment, while at higher levels cultures, social norms, and political paradigms influence the way societies interact with nature.[4]

Industrial societies have tended to view their natural resource base, including water, as an asset to utilize for social and economic growth and technological innovation. The emergence of global environmental problems has increased awareness about the interconnectedness of human society and the physical natural environment.[5] Many sectors of post-industrial society have subsequently realized that the unregulated exploitation of the natural resource base endangers not only nature but the human society that depends upon it. The diverse underlying beliefs and attitudes about the environment and its function have motivated and provided fuel for US environmental movements over the past century.

US Environmental Movements

Shifts in environmental beliefs and actions are represented by four environmental movements: preservation, conservation, contemporary environmentalism, and grassroots environmental management.[6] Each of these movements has corresponding basic beliefs about the function of nature and how humans interact with their

---

[2] Ajzen, Icek. 2001. "Nature and Operation of Attitudes." *Annual Review of Psychology* 52:27–58.

[3] Buttel, Fredrick H. and Craig R. Humphrey. 2002. "Sociological Theory and the Natural Environment." Chapter 2 in handbook of *Environmental Sociology* edited by Riley E. Dunlap and William Michelson, Westport, CT: Greenwood Press; Weber, Edward P. 2000. "A New Vanguard for the Environment: Grass-Roots Ecosystem Management as a New Environmental Movement." *Society and Natural Resources* 13:237–259.

[4] Lundmark, Carina. 2007. "The New Ecological Paradigm Revisited: Anchoring the NEP Scale in Environmental Ethics." *Environmental Education Research* 13:329–347.

[5] Stern, Paul C., Oran R. Young, and Daniel Druckman. 1992. *Global Environmental Change: Understanding the Human Dimensions.* Washington, DC: National Academy Press.

[6] Weber, Edward P. 2000. "A New Vanguard for the Environment: Grass-Roots Ecosystem Management as a New Environmental Movement." *Society and Natural Resources* 13:237–259.

natural environment. During the *preservation movement* of the late 1800s, the belief was that nature is something to be valued for its aesthetic beauty and as a result public lands were set aside to preserve wilderness from industrialization and civilization.[7] Dominant beliefs of the *conservation movement* were centered on nature existing to benefit humankind; that natural resources are of greatest value when extracted, and that science and technologies can be used to efficiently extract and manage natural resources.[8]

The fragility of nature and the impacts of human society moved to the forefront of social discourse as pollution of water and air by industries threatened the quality of life in communities where they were located. The *contemporary environmental movement* launched a series of legislative, public monitoring, and regulatory initiatives to restrict exploitation of natural resources, control environmental degradation, and protect the environment.[9] Groups with concern about human impacts on the environment pushed passage of a series of environment-related laws and legislation – the 1970 National Environmental Policy Act, the 1970 Clean Air Act, the 1972 Water Pollution Control Act (Clean Water Act), and the 1976 Toxic Substances Control Act. Toward the end of the 1970s, two major US environmental disasters, the Three Mile Island nuclear accident and the toxic contamination of Love Canal further engaged the public for environmental action.[10] The single theme during this period was that nature not only had intrinsic worth apart from human beings, it had the same legal rights to protection as humans.[6]

A fourth movement, *grassroots environmental management*, merges an awareness of the fragility of the environment with concrete conservation and practical land use interventions.[6] Participants are environmentally active but do not consider themselves environmental activists. This grassroots movement is illustrated by the development of local watershed groups across the USA with less interest in politicizing the environment and more interest in engaging in activities that solve the problem of water quality and quantity. The legacies of these four movements and their public beliefs and actions continue to influence public policies and decisions.

---

[7] Switzer, Jacqueline Vaughan. 1997. *Green Backlash: The History and Politics of Environmental Opposition in the U.S.* Boulder, CO: Lynne Rienner.

[8] Cawley, R. McGreggor. 1993. *Federal Land, Western Anger: The Sagebrush Rebellion and Environmental Politics.* Lawrence, KS: University Press of Kansas; Weber, Edward P. 2000. "A New Vanguard for the Environment: Grass-Roots Ecosystem Management as a New Environmental Movement." *Society and Natural Resources* 13:237–259.

[9] Mitchell, Robert Cameron. 1991. "From Conservation to Environmental Movement: The Development of the Modern Environmental Lobbies." Pp. 81–114 in *Government and Environmental Politics* edited by M. J. Lacey, Baltimore, MD: Johns Hopkins University Press.

[10] Taylor, Dorceta E. 2005. "American Environmentalism: The Role of Race, Class and Gender in Shaping Activism 1820–1995." Pp. 87–106 in *Environmental Sociology: From Analysis to Action* edited by Leslie King and Deborah McCarthy. Lanham, MD: Rowman & Littlefield Publishers, Inc.; Schnaiberg, A. 2001. "Environmental Movements since Love Canal: hope, despair and [im] mobilization?" *Buffalo Environmental Law Journal* 8:256–269.

Environmental Experiences and Concerns

Environmental concern is a phrase broadly used to refer to the degree that a person recognizes environmental problems and indicates a willingness to contribute to solving those problems.[11] Common factors used to explain variations in environmental concern include age, sex, education, income, occupation, residence (rural vs. urban), political party affiliation, and political ideology.[12] Over the years, the predictor variable list has expanded to include race, ethnicity, geographic region, farm vs. non-farm residence, occupations, and socialization.[13] Environmental concerns are thought by some scholars to be linked to perceived conditions and direct experience with or exposure to pollution.[14] Those who depend directly on the use

---

[11]Dillman, Don A. and James A. Christenson. 1972. "The Public Value for Pollution Control." Pp. 237–256 in *Social Behavior, Natural Resources and the Environment* edited by William R. Burch, Jr., Neil H. Creek, Jr., and Lee Taylor. New York: Harper and Row; Buttel, Frederick H. and William L. Flinn. 1976. "Economic Growth Versus the Environment: Survey Evidence." *Social Science Quarterly* 57:410–420; Riley E. Dunlap and Jones, Robert Emmet. 2002. "Environmental Concern: Conceptual and Measurement Issues." In *Handbook of Environmental Sociology* edited by R. E. Dunlap and W. Michelson, Westport, CT: Greenwood Press.

[12]Van Liere, Kent D. and Riley E. Dunlap. 1980. "The Social Bases of Environmental Concern: A Review of Hypotheses, Explanations, and Empirical Evidence." *The Public Opinion Quarterly* 44:181–197.

[13]Tremblay, Kenneth R. and Riley E. Dunlap. 1978. "Rural Urban Residence and Concern with Environmental Quality: A Replication and Extension." *Rural Sociology* 43:474–491; Buttel, Frederick H., Gilbert W. Gillespie, Jr., Oscar W. Larson III, and Craig K. Harris. 1981. "The Social Bases of Agrarian Environmentalism: A Comparative Analysis of New York and Michigan Farm Operators." *Rural Sociology* 46(3):391–410; Lowe, George D. and Thomas K. Pinhey. 1982. "Rural-Urban Differences in Support for Environmental Protection." *Rural Sociology* 47:114–128; Hand Carl M. and Kent Van Liere. 1984. "Religion, Mastery-Over-Nature, and Environmental Concern." *Social Forces* (2):555–570; Freudenburg, William R. 1991. "Rural-Urban Differences in Environmental Concern: A Closer Look." *Sociological Inquiry* 61(2):167–198; Jones, Robert Emmet and Riley E. Dunlap. 1992. "The Social Bases of Environmental Concern: Have They Changed Over Time?" *Rural Sociology* 57:28–47; Mohai, Paul and Bunyan Bryant. 1998. "Is There a 'Race' Effect on Concern for Environmental Quality?" *The Public Opinion Quarterly* 62:475–505; Nooney, Jennifer G., Eric Woodrum, Thomas J. Hoban, and William B. Clifford. 2003. "Environmental Worldview and Behavior: Consequences of Dimensionality in a Survey of North Carolinians." *Environment and Behavior* 35(6):763–783; Johnson, Cassandra Y., J. Michael Bowker, and H. Ken Cordell. 2004. "Ethnic Variation in Environmental Belief and Behavior: An Examination of the New Ecological Paradigm in a Social Psychological Context." *Environment and Behavior* 36(2):157–186; Greenberg, Michael R. 2005. "Concern About Environmental Pollution: How Much Difference Do Race and Ethnicity Make? A New Jersey Case Study." *Environmental Health Perspectives* 113:369–374; Milfont, Taciano L., John Duckitt, and Linda D. Cameron. 2006. "A Cross-cultural Study of Environmental Motive Concerns and Their Implications for Proenvironmental Behavior." *Environment and Behavior* 38(6):745–767.

[14]Lowe, George D. and Thomas K. Pinhey. 1982. "Rural-Urban Differences in Support for Environmental Protection." *Rural Sociology* 47:114–128; Tremblay, Kenneth R. and Riley E. Dunlap. 1978. "Rural Urban Residence and Concern with Environmental Quality: A Replication and Extension." *Rural Sociology* 43:474–491; Van Liere, Kent D. and Riley E. Dunlap. 1980. "The Social Bases of Environmental Concern: A Review of Hypotheses, Explanations, and Empirical Evidence." *The Public Opinion Quarterly* 44:181–97.

of natural resources for their livelihoods have been reported as more likely to have lower levels of environmental concern.[15]

Results from extensive research show that the associations found between environmental concern and these sociodemographic variables are not consistent. However, there is a general pattern that younger adults, the highly educated, and political liberals usually express greater concern about the environment than do their respective counterparts.[16] Beyond the framework of sociodemographic variables, some studies looked at factors such as community attachment,[17] general beliefs,[18] the presence and severity of specific environmental problems,[19] and knowledge and awareness of consequences societies have on the environment.[20]

We examined how environmental beliefs and attitudes are related to concerns about water quality issues – specifically, how a person's general environmental views are associated with the person's perceptions of water quality, willingness to learn about water quality issues, and change of behavior regarding water.

A Multi-State Water Issue Survey

Our study uses data from a multi-state water issue survey completed in 36 of the US states (2002–2009) conducted by Dr. Robert Mahler, University of Idaho under a USDA Integrated Water Quality project.[21] Households were randomly sampled from phone books in each state, and calculation of targeted sample size was based on the total population of the state. Mailed surveys were sent to the sampled names and addresses, but any adult in the household, whether or not the addressee of the mailed survey could fill out the survey questionnaire. Questions and wordings in the surveys

---

[15]Freudenburg, William R. 1991. "Rural-Urban Differences in Environmental Concern: A Closer Look." *Sociological Inquiry* 61(2):167–198.

[16]Xiao, Chenyang and Aaron M. McCright. 2007. "Environmental Concern and Sociodemographic Variables: A Study of Statistical Models." *The Journal of Environmental Education* 38(2):3–13.

[17]Vorkinn, Morit and Hanne Riese. 2001. "Environmental Concern in a Local Context: The Significance of Place Attachment." *Environment and Behavior* 33(2):249–263; Brehm, John M., Brain W. Eisenhauer, and Richard S. Krannich. 2006. "Community Attachments as Predictors of Local Environmental Concern: The Case for Multiple Dimensions of Attachment." *American Behavioral Scientist* 50(2):142–165.

[18]Olofsson, Anna, and Susanna Ohman. 2006. "General Beliefs and Environmental Concern: Transatlantic Comparisons." *Environment and Behavior* 38:768–790.

[19]Arcury, Thomas A. and Eric H. Christianson. 1990. "Environmental Worldview in Response to Environmental Problems." *Environment and Behavior* 22(3):387–407.

[20]Dunlap, Riley. 1998. "Lay Perceptions of Global Risk: Public Views of Global Warming in Cross-National Context." *International Sociology* 13(4):473–498; Hayes, Bernadette. 2001. "Gender, Scientific Knowledge and Attitudes Toward the Environment: A Cross-National Analysis." *Political Research Quarterly* 54:657–671.

[21]This survey is part of a national project conducted by Dr. Robert Mahler, Professor of Soil and Environmental Sciences at University of Idaho under USDA CSREES project 2004-51130-02245.

varied from state to state, but the total survey length was approximately 50 questions. There were several core common questions across states that asked about respondents' perceptions of water quality, water use importance, factors responsible for water pollution, sources of information about water, general environmental attitudes, and demographic information. Standard mail survey methods as recommended by Dillman[22] were followed in each of the states, with a total of 9,332 completed surveys, and response rates ranging from 37 (Massachusetts) to 70% (Wyoming).

The US Environmental Protection Agency (EPA) groups the 50 states into ten regions based on geography and regional conditions. We examine core survey items held in common by the individual states as well as aggregate our findings by EPA region. The water survey data cover states in eight of the ten EPA regions, excluding the Great Lakes Region (Region 5) and the Southeast Region (Region 4, with surveys in Tennessee but not other states).

Core Survey Items

General environmental attitude, perceptions of ground and surface water quality, willingness to learn more about water issues, and changes in land use practices that affect water were core indicators across all states. The general environmental attitude indicator attempts to capture the range of attitudes about the function of the environment from human use only to total environment protection without regard to use and function. Due to survey questionnaire length limitations, general environmental attitude is measured as a single item rather than multiple items as in the New Ecological Paradigm scale.[23] However, the underlying rationale is similar – to capture the extent to which a person has a pro-anthropocentric (people-centered) versus a pro-ecocentric (pro-environmental) worldview. Respondents were asked to indicate where they stand on environmental issues by placing a mark on a line with numbers 1–10, where 1 represents support for total natural resource use and 10 represents support for total environmental protection, with the median point (5.5) representing an equally distant position between the two.

Two perceptions of water quality questions measure respondent opinions about the condition of their surface and ground waters. Response options were 1 = poor, 2 = fair, and 3 = good or excellent. For measures of willingness to learn, respondents were asked, "Would you like to learn more about any of the following water quality areas?" Although six[24] of the regions had this question on their surveys, the options

[22]Dillman, Don A. 2000. *Mail and Internet Surveys,* 2nd ed. New York: Wiley.

[23]Dunlap, Riley E., Kent Van Liere, Angela G. Mertig, and Robert Emmet Jones. 2000. "New Trends in Measuring Environmental Attitudes: Measuring Endorsement of the New Ecological Paradigm: A Revised NEP Scale." *Journal of Social Issues* 56 (3):425–442.

[24]Region 1 (ME, VT, NH, MA, RI, CT), Region 2 (NY, NJ), Region 3 (PA, WV, VA, DE, MD, DC), Region 4 (NC, SC, GA, KY, TN, MS, AL, FL), Region 6 (AR, LA, OK, TX, NM), Region 7 (IA, MO, KS, NE).

were not the same. Therefore, only nine identical items[25] are used in this analysis and summed to make a "learn" index with respondents marking two items on average.

Change of behavior was measured by a set of questions asking "Have you or someone in your household done any of the following as part of an individual or community effort to conserve water or preserve water quality?" Respondents had the option to check "changed the way your yard is landscaped," "changed how often you water your yard," and "changed your use of pesticides, fertilizers or other chemicals." Six regional surveys included the behavior change questions.[24] When these six regions are analyzed, the total number of responses is reduced to 5,508.

Perceptions of Water Quality, Environmental Attitudes, and Willingness to Learn
Surface and Ground Water Quality Perceptions

Actual water quality conditions vary considerably across US states and regions.[26] Documentation of the number of impaired water bodies and type of impairment in each state is an ongoing process carried out by state agencies. The EPA public web site lists the number of impaired waters by state. States with many rivers and lakes have documented many impaired waters: Pennsylvania (6,957 in 2004), New Hampshire (5,192 in 2004), California (686 in 2002), New York (792), and Ohio (428 in 2004). States with fewer water bodies also report impairments: Utah (141 in 2004), Arizona (68 in 2004), and Wyoming (129 in 2004). While these data are public, citizen awareness and perceptions of impairment vary across states and regions. Maps 8.1 (ground water) and 8.2 (surface water) offer a spatial view of state residents' general perceptions of their surface and ground water quality.

The overall average perception in all 36 states for surface water quality is 2.0 or fair. The overall average perception of ground water quality is 2.4, about half way between fair and good/excellent, and higher than that of surface water. Some states, like Alaska (2.7 surface; 2.7 ground), Maine (2.5 surface; 2.5 ground), and Vermont (2.5 surface; 2.7 ground), consistently rate their water quality higher than the national average perception scores on both surface and ground water. Residents of Nevada (1.7 surface; 1.9 ground) and Delaware (1.6 surface; 1.9 ground) rate both ground and surface water quality lower than average. Residents in some states on average rate ground water high but surface water lower (Hawaii 2.6 ground, 1.8 surface; Iowa 2.5 ground, 1.8 surface; Missouri 2.5 ground, 1.8 surface). Generally speaking, within each state, ground water quality is perceived to be better than surface water quality. Although it might be the case that compared with surface water, ground water is less

---

[25] The nine items to make the "learn" index were watershed management, watershed/environmental restoration, irrigation management, animal manure and waste management, nutrients and pesticide management, private well and septic management, public drinking water and human health, water policy and economics, and home and garden landscaping.

[26] National Water Quality Inventory. (http://www.epa.gov/305b) Retrieved 7-3-07.

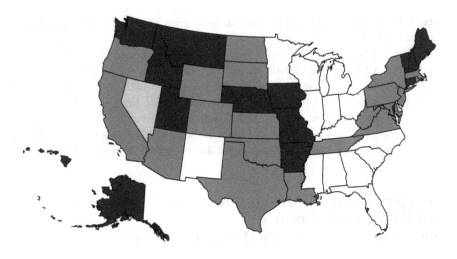

■ 2.5 to 3.0
▨ 2.0 to 2.4
▢ Less than 2.0
☐ No data available

Map 8.1 What is the quality of ground water in your area? (1=poor; 2=fair; 3=good or excellent)

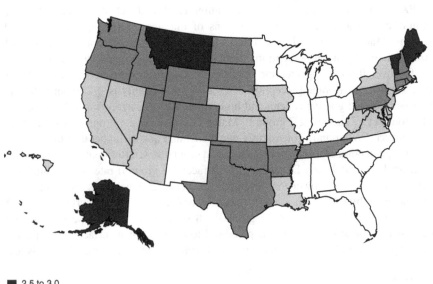

■ 2.5 to 3.0
▨ 2.0 to 2.4
▢ Less than 2.0
☐ No data available

Map 8.2 What is the quality of surface water in your area? (1=poor; 2=fair; 3=good or excellent)

polluted and therefore has better quality, our data do not tell us about actual conditions. There are a variety of factors that influence public water quality perceptions. For example, perceptions of surface water quality could be formed from visual observation and direct experiences with rivers, lakes, and streams such as beach advisories for high bacteria, muddy water, and algae blooms. Ground water is less visible and perceptions of quality are more likely to be formed through indirect sources like media reports, word of mouth, and public information programs.

Environmental Attitudes

An examination of our data on environmental attitudes suggests that on average, US citizens tend to hold the middle ground between anthropocentric and ecocentric views. The variations across states and regions in environmental attitudes are shown in Table 8.1. The overall mean score is 5.8 on the continuum between 1 and 10.

A comparison of state mean scores shows that eastern states have attitude scores that lean toward protecting the environment for its own value (Vermont 6.4; Delaware 6.4; New York 6.4; Rhode Island 6.3). Mid-western and western states with strong agriculture, mining, and forestry landscapes have environmental attitudes that reflect a tilt toward functional use viewpoints (North Dakota 5.1; Oklahoma 5.3; Wyoming 5.3; and Idaho 5.4).

Willingness to Learn

Willingness to learn about water issues (Fig. 8.1) and behavioral changes to protect water resources (Fig. 8.2) are examined by region, because states within the same region tend to show homogeneous patterns. In general, respondents in all six regions report the highest willingness to learn about public drinking water and human health. The percent of willingness to learn about this issue ranges from 38.2% in Region 1 to 53% in Region 7. A second strong interest in learning focuses on home and garden landscaping. This topic has both rural and urban applications as well as low regional variation.

In comparison, irrigation – an agricultural practice – received the lowest percentage of willingness to learn. In Regions 1–4, only 6.2–8.8% of respondents said they would be willing to learn about irrigation issues. In Regions 6 and 7, however, a higher percentage (13 and 10%, respectively) of the respondents indicated a willingness to learn about irrigation. This outcome is likely because irrigation is associated with low annual rainfall locations in central and western states compared with eastern states with higher rainfall and less need to irrigate for agricultural production. Approximately 20–30% of respondents indicated a willingness to learn about the other seven water issues.

Changing Practices

Twenty to 25% of respondents in the six regions said they have changed the way their yard is landscaped (Fig. 8.2). Forty-nine to fifty-nine percent of all respondents

Table 8.1 Environmental attitudes

State	Mean	Rank	Region
Vermont	6.42	1	1
Delaware	6.40	2	3
New York	6.39	3	2
Hawaii	6.38	4	9
Rhode Island	6.32	5	1
New Jersey	6.27	6	2
Maryland	6.21	7	3
California	6.19	8	9
Massachusetts	6.13	9	1
Colorado	6.13	10	8
Connecticut	6.10	11	1
Virginia	6.03	12	3
New Hampshire	6.03	13	1
Montana	5.84	14	8
Pennsylvania	5.83	15	3
Missouri	5.79	16	7
West Virginia	5.75	17	3
Washington	5.74	18	10
Louisiana	5.72	19	6
Oregon	5.70	20	10
Tennessee	5.67	21	4
Arizona	5.66	22	9
Texas	5.66	23	6
Kansas	5.65	24	7
Nevada	5.64	25	9
Iowa	5.64	26	7
Maine	5.63	27	1
Nebraska	5.50	28	7
South Dakota	5.50	29	8
Arkansas	5.41	30	6
Alaska	5.39	31	10
Idaho	5.38	32	10
Utah	5.35	33	8
Wyoming	5.33	34	8
Oklahoma	5.31	35	6
North Dakota	5.14	36	8
Overall mean	5.84		

reported they have changed how often they watered their yard. There is a geographic pattern from east to west, with a higher proportion of respondents changing watering practices to preserve water in western regions corresponding with reductions in annual rainfall. The percent of respondents who have changed their use of pesticides, fertilizers, or other chemicals ranged from 28% in Region 6 (AK, LA, OK, and TX) to 43% in Region 1 (ME, VT, NH, MA, RI, and CT).

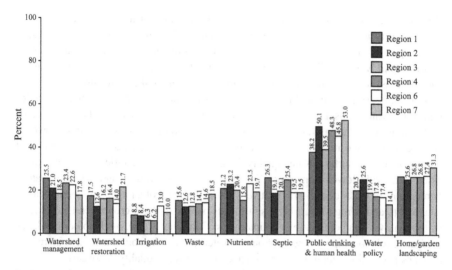

Fig. 8.1 Willingness to learn by region

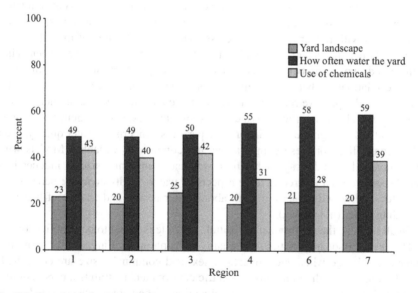

Fig. 8.2 Reported behavioral change

Correlations and Causal Models

We developed several models using available data from the survey to discover relationships among willingness to learn, perceptions of water quality, changes in behaviors, and environmental attitudes controlling for demographic characteristics (age, education, sex, and community size). We hypothesized that there would be a significant association between general environmental attitudes and concern for water quality issues net of demographic variables, because water quality is a specific aspect of the environment. A positive relationship between environmental attitudes and water quality concern would mean that a more pro-ecocentric attitude would be associated with lower perceptions of water quality. We further proposed that pro-ecocentric attitude, higher willingness to learn, and more changed behavior regarding water conservation and water quality preservation would be associated.

Bivariate correlations (not controlling for demographic characteristics) showed a significant negative relationship between environmental attitudes and perceptions of both ground and surface water quality. Thus, more pro-ecocentric views tend to be associated with lower confidence in the quality of ground and surface waters. Perceptions of water quality are positively associated with older age, higher educational attainment, and being male. This result means people in these groups tend to perceive water quality as higher than their counterparts. Community size, on the other hand, has a negative relationship with perceptions of water quality. As community size increases, people's perceptions about their ground and surface water quality decreases.

More pro-ecocentric environmental views tend to vary together with a greater willingness to learn about water issues.[27] A high pro-ecocentric environmental attitude is also positively associated with three changes of behaviors: changes in landscaping their yard, watering their garden, or the use of chemicals. Further, respondents with higher education are associated with all three of these behavior changes. Other significant correlations include a positive association between being female and two of the behavior changes – watering yard and use of chemicals; a positive relationship between community size and changed frequency of garden watering; and a positive relationship between age and changed use of chemicals.

Given the above significant associations, one may expect that general environmental attitudes might be a predictor for a person's concern for water quality, even when controlling for these demographic and community variables. Models with environmental attitudes, age, education, sex, and community size as independent variables were used to predict three aspects of water quality concern: perceptions of water quality, willingness to learn about water issues, and change of behavior regarding water (not shown).

In all models, the environmental attitude variable is found to be significantly (at a 0.05 level) associated with water quality perceptions, willingness to learn, and behavioral change when age, education, sex, and community size are controlled. This finding means the more pro-ecocentric environmental attitude a person holds, the more likely they will have a lower perception of their surface and ground water quality, regardless of their demographic characteristics. Further, the person with a stronger pro-ecocentric environmental attitude will be more likely to indicate a willingness to learn more about water issues and have changed behaviors regardless of their demographic characteristics. The causal models confirm our hypotheses about the relationships between general environmental attitudes and water quality

---

[27] All of these bivariate associations are found to be statistically significant (at .05 level).

concern. However, these are weak models explaining less than 5% of the variation. Our data set is inadequate to explain factors that are influencing attitudes and changes in behavior. There are clearly other more important factors that are influencing willingness to learn and changes in water practices. Future research should include biophysical conditions, exposure to media reports, distance from a water body, and additional attitude and belief measures to build more robust models.

Conclusion

Water is a critical environmental issue facing societies around the world. How people perceive their environment and the effects of drought, flooding, and water contamination will influence their willingness to invest public and private resources to address this increasingly complex problem. Our research shows that although there are significant correlations between water quality concern and general environmental attitudes, a causal model of environmental attitudes, demographical characteristics, and community size does not explain very much of the variation in water quality concern. This finding is a reminder of the complexity of water issues, which involve attitudes, actual conditions, social organization, and cultural norms. Further, these factors may be different for residents in each of the 36 states and are likely associated with actual water conditions in the state as well as public policies and the infrastructure in place to address water issues.

Given these regional and state differences, multilevel models with variables that focus on social context (e.g., regional differences in geography, culture, and economy) might work better in explaining individual environmental concern. Findings about residents' willingness to learn about water, their change of behaviors, and comparisons of residents' perceptions about water quality in these states with the actual water conditions will help project designers and practitioners learn from the local knowledge of the specific area. This process can lead to better-designed programs that educate citizens about their water and engage them in local watershed management efforts.

Chapter 9
Communities of Interest and the Negotiation of Watershed Management

Max J. Pfeffer and Linda P. Wagenet

The Legal Equivalent of a Hoover Dam

The New York City Watershed Memorandum of Agreement (MOA) signed in January 1997 was an extraordinary accomplishment. Some have called this unprecedented agreement "the legal equivalent of a Hoover Dam."[1] The MOA represents a special kind of accomplishment in community development – the creation of a "watershed community of interest." This community is described in the MOA as "shar[ing] the common goal of protecting and enhancing the environmental integrity of the Watershed and the social and economic vitality of the Watershed communities."[2] The MOA was signed by approximately 40 upstate towns and villages, environmental groups, the United States Environmental Protection Agency (EPA), New York State, and New York City (NYC). The agreement serves as a blueprint for NYC's watershed management strategy for water sources west of the Hudson River.[3] It cost approximately one billion dollars over 10 years. Figure 9.1 displays a map of the watershed and highlights some of its prominent features.

The significant lesson to be drawn from the NYC case is that communities can more effectively respond to environmental problems by collaborating with other communities. But human settlements are not typically organized according to

---

[1] Platt, Rutherford H., Paul K. Barten, and Max J. Pfeffer. 2000. "A Full, Clean Glass? Managing New York City's Watersheds." *Environment* 42(5):9–20.

[2] New York City Watershed Memorandum of Agreement (MOA). 1997. January (unpublished manuscript), Article IV, Para. 97.

[3] This analysis is limited to the portion of the watershed found west of the Hudson River. NYC harvests 90% of its water supply in this area by means of a set of constructed reservoirs and controlled lakes. NRC (National Research Council). 2000. *Watershed Management for Potable Water Supply: Assessing the New York City Strategy*. Washington, DC: National Academy Press.

M.J. Pfeffer (✉)
Department of Development Sociology, Cornell University, 267 Roberts Hall, Ithaca, NY 14853, USA
e-mail: mjp5@cornell.edu

L.W. Morton and S.S. Brown (eds.), *Pathways for Getting to Better Water Quality: The Citizen Effect*, DOI 10.1007/978-1-4419-7282-8_9, © Springer Science+Business Media, LLC 2011

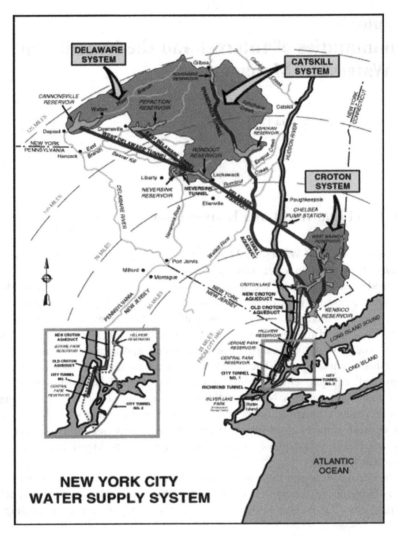

Fig. 9.1 Map of the New York City watershed (The Delaware, Catskill and Croton Systems are shown as shaded areas from which water is collected and in which land use and other watershed protection measures apply)

watershed boundaries, and, in large watersheds such as the Delaware and Hudson River basins, the lack of unifying organizations among politically and spatially disconnected communities makes coordinated watershed actions especially difficult.[4] One of the most significant challenges to achieving collaboration among

---

[4]Pfeffer, Max J. 2003. "The Watershed as Community." In *Encyclopedia of Community: From the Village to the Virtual World* edited by Karen Christensen and David Levinson, Thousand Oaks, CA: Sage.

watershed communities is the development of a widely shared set of watershed management objectives. We refer to individuals and communities with shared watershed management objectives as a *community of interest*.

Forging communities of interest among watershed groups is a necessary and powerful basis for successful negotiation of diverse interests into acceptable watershed management plans. The realization of such shared objectives demands active stakeholder engagement. Indeed, emphasis on stakeholder engagement in watershed management has come to be widely acknowledged as essential for successful watershed management.[5] Given the variety of social, economic, and ecological interests typically found across a watershed, several developments must take place for a community of interest to form. Three are of particular importance: (1) acceptance of common terms of reference, (2) agreement on physical and legal boundaries, and (3) establishment of standards of equity.

Common Terms of Reference

Reaching agreement on common terms of reference is necessary to resolve conflicts and establish a widely shared vision of the common good. There must be considerable give and take between parties, and, in this process of give and take, terminology and meanings evolve in ways that express the common interests of the parties involved. This change reflects a shift in the nature of the social relationships between communities, from an emphasis on excluding the demands of others to one of greater reciprocity among the parties involved.[6]

---

[5]Wagenet, Linda and Max J. Pfeffer. 2007. "Organizing Citizen Engagement for Democratic Environmental Planning." *Society and Natural Resources* 20(9):801–813.

[6]In more academic terms, we refer here to a shift from exclusionary social relations, or bonding social capital, to reciprocal ones, or bridging social capital. Social capital is a conceptual cornerstone in understanding the organizational foundations for watershed management. We use "social capital" to refer to the value or utility of social relationships in achieving a desired outcome. This definition is consistent with the foundational literature that treats social capital as a means of gaining access to economic resources (Portes 1998; Bourdieu 1986), but opens the possibility that social capital can be used to pursue ends not purely economic in nature. By *community* social capital, we mean social relations between communities. While individuals are involved in these relations, we are interested in their interactions as agents representing communities. Following Meyer and Jesperson (2000:101), we conceive of "agency" as "legitimated representation of some legitimate principle, which may be an individual [or] an organization, a nation state, or abstract principles (like those of science ...)." (Portes, Alejandro. 1998. "Social Capital: Its Origins and Applications in Modern Sociology." *Annual Review of Sociology* 24:1–12; Bourdieu, Pierre. 1986. "The Forms of Capital." Pp. 241–258 in *Handbook of Theory and Research for the Sociology of Education,* edited by J. D. Richardson, Westport, CT: Greenwood Press; Meyer, John W. and Ronald L. Jepperson. 2000. "The 'Actors' of Modern Society: The Cultural Construction of Social Agency." *Sociological Theory* 18(1):100–120).

Boundary Conditions

Physical or legal boundaries on access to and legitimate use of land and other natural resources affect property owners differently. New rules and regulations create uncertainty among landowners about development rights on land already owned. Landowners in particular face the risk of financial loss. To ensure stability and confidence of the watershed population, there needs to be clarity about physical and legal boundaries, and these boundaries must be agreed upon by all parties involved.

Fairness

The articulation of grievances and demands for equitable treatment are an important part of the process of forming a community of interest. Controversy and uncertainty heighten awareness of certain forms of inequality and injustice, and lead to the articulation of a clear set of demands that establish a basis for resolving differences.[7]

Community organizations play a key role in the creation of common terms of reference, accepted boundary conditions, and mechanisms to assure fairness. To illustrate this point, we review the role of key organizations in the development of the historic New York City watershed agreement.

The New York City Watershed

The collaborative NYC Watershed Memorandum of Agreement was not necessary or even expected in 1990 when the NYC Department of Environmental Protection (DEP) produced its *Discussion Draft of Proposed Regulations for the Protection from Contamination, Degradation, and Pollution of the New York City Water Supply and its Sources.*[8] The revised Rules and Regulations would have restricted a variety of developments. In addition, the *Draft Generic Environmental Impact Statement*, prepared in response to the proposed regulatory revisions, included a proposal for NYC to acquire watershed land. In late 1993, NYC filed a state application for a water supply permit including plans to acquire 10,000 acres in the watershed and submitted to EPA a Long-Term Watershed Protection and Filtration Avoidance Program for the Catskill/Delaware System. Uncertainty over NYC's intent to use eminent domain to gain control of land and the perception that NYC was shifting

---

[7]For an example of the formation of a community of interest – Pfeffer, Max J., John W. Schelhas and Leyla Ann Day. 2001b. "Forest Conservation, Value Conflict, and Interest Formation in a Honduran National Park." *Rural Sociology* 66(3):382–402.

[8]New York City Department of Environmental Protection (DEP) 1990. *Discussion Draft of Proposed Regulations for the Protection from Contamination, Degradation and Pollution of the New York City Water Supply and Its Sources*. Corona, NY: NYCDEP.

the costs of watershed protection to upstate communities resulted in the deterioration of already strained relations between NYC and upstate communities.

NYC's proposed actions aroused strong opposition from watershed residents who feared that economic development would be stifled, property values would drop, and the local tax base would be eroded.[9] However, lack of cohesiveness among watershed communities limited the setting and pursuing of common goals despite shared fears. This limitation prevented the negotiation of a partnership agreement with NYC. Opposition from within the watershed was sustained by intense loyalties among members of the individual communities who were intent on protecting local interests. The strong local identity helped people preserve a clear sense of community, but tended to exclude relations with those from other communities. The result was development of a *culture of resistance*.

Rural, west-of-Hudson communities embodied this culture of resistance, which was based on shared knowledge about the watershed and common values, norms, and attitudes about their local community and environment. Throughout the watershed, the culture defined NYC's treatment of watershed communities as oppressive, especially in light of NYC's treatment of the rural towns during construction of the water system in the early twentieth century. Local residents regularly reaffirmed this view in their depictions of the relationship between the rural communities and NYC. One Catskill area newspaper represented the attitude toward NYC when it wrote, "Upstate residents cannot rely on the city's willingness to abide by any statements or principle of law regarding its responsibilities in the watershed."[10]

Although watershed communities shared this culture of resistance, population trends and home rule, or local self-governance, reinforced community insularity and fragmented interests within the watershed. There are 40 towns west of the Hudson River with some land in the NYC Watershed, and they experienced relatively little population growth since the early twentieth century. There are few opportunities for newcomers entering the region. As one resident of a watershed community noted, "One of the largest exports from Delaware County is educated children, and it has been that way for fifty years."[11] In the absence of newcomers, local communities were more homogeneous, and the culture of resistance and exclusion was reinforced.

Adherence to home rule as a principle of land use management also reinforced community insularity as well as solidarity. Local control over land use is a central

---

[9]Finnegan, Michael C. 1997. New York City's Watershed Agreement: A Lesson in Sharing Responsibility. *Pace Environmental Law Review* 14:577–644; Schneeweiss, Jonathan. 1997. "Watershed Protection Strategies: A Case Study of the New York City Watershed in Light of the 1996 Amendments to the Safe Drinking Water Act." *Villanova Environmental Law Journal* 9:77–119.

[10]We monitored six local newspapers' coverage of watershed issues for approximately 6 years beginning in 1994. Newspapers monitored included the Catskill Mountain News (CMN), Daily Freeman (DF), Delaware County Times (DCT), Deposit Courier (DC), New York Times (NYT), and the North Country News (NCN).

[11]Stave, Krystyna Anne. 1998. *Water, Land, and People: The Social Ecology of Conflict over New York City's Watershed Protection Efforts in the Catskill Mountain Region.* PhD Dissertation, NY: Yale University, p. 253.

element of home rule, "the right to self-government in local affairs."[12] Catskill towns have elected Town Supervisors and Town Boards. Within towns, incorporated villages may also have their own village governments. Planning boards are found in some but not all towns. "'Home rule,' a town supervisor explained, 'means that every town is different.'"[13]

For the NYC Watershed Agreement to come about, a new, more inclusive identity had to form, one that would encompass relations with other communities and NYC. Sharing this identity, individual watershed communities would see the formation of linkages with other communities as an asset. An upstate/downstate partnership would be seen as an opportunity rather than as a threat. This identity would have to encompass a variety of interests, and the expectation would need to be that competing interests could be reconciled through some form of reciprocity.

A Watershed Identity

The negotiation of the NYC Watershed Agreement can be viewed as a "community development project." It involved the development of a watershed identity that encompassed the interests of both water quality protection for downstream consumers and social and economic well-being for upstream residents. However, the watershed community of interests could not be built without several organizational developments occurring. Two organizations in particular played essential roles: the Watershed Agricultural Council and the Coalition of Watershed Towns. These organizations were able to help negotiate the three necessary conditions: agreed upon terms of reference, boundary conditions, and a fair solution.

The Watershed Agricultural Council (WAC) was created in 1990 by an Ad Hoc Task Force on Agriculture and New York City Watershed Regulations. The WAC was to be a "city–farm partnership" made up of governmental agencies (local, state, and NYC) and the farm community. The WAC model of upstate/downstate partnership paved the way for a broader watershed community of interests. Understanding why the breakthrough occurred in agriculture first helps us understand some of the key organizational ingredients that supported the creation of a community of interest on a broader scale.

Agriculture already had a strong community of interest among organizations that typically work across township and county boundaries, thereby providing a social bridge between political units. These organizations, which have strong ties to local farmers, include the county Soil and Water Conservation Districts (SWCD) and the

---

[12]Nolon, John R. 1993. "The Erosion of Home Rule through the Emergence of State Interests in Land Use Control." *Pace Environmental Law Review* 10(2):497–562, p. 55.

[13]Stave, Krystyna Anne. 1998. *Water, Land, and People: The Social Ecology of Conflict over New York City's Watershed Protection Efforts in the Catskill Mountain Region.* PhD Dissertation, NY: Yale University, p. 253.

New York State Department of Agriculture and Markets, Cornell Cooperative Extension, and organizations such as Farm Bureau. Many members of the farm community, including these local, state, and federal agricultural agencies, believed that the proposed regulations on agriculture threatened the continued viability of farms in the New York City watersheds, especially dairy and livestock farms.[14] Strong farmer opposition to the proposed regulations was expressed in a series of meetings organized by the Farm Bureau and the State Soil and Water Conservation Committee. The uproar raised by the farm community cast a long, dark shadow over the proposed regulations.

Efforts to Balance Competing Needs

In December 1990, the New York State Department of Agriculture and Markets and the NYC Department of Environmental Protection (DEP) jointly convened an Ad Hoc Task Force on Agriculture and New York City Watershed Regulations. This group's recommendations were the basis for the first organizational effort to balance drinking water quality protection with local control over land use. The Ad Hoc Task Force published its "Policy Group Recommendations" in December 1991, and NYC formally accepted them. NYC agreed to withdraw the proposed regulations that applied to agriculture if farmers adopted the voluntary Whole Farm Planning Program recommended by the Ad Hoc Task Force. The recommendations also clearly established the leading role of agriculture by declaring it "the preferred land use" in the watershed, and acknowledged the importance of maintaining the profitability of the area's farms.[15] In establishing the central importance of agriculture within the watershed, the Ad Hoc Task Force established many precedents that would later more generally inform management of the NYC watershed.[16] A key development toward eventual agreement was the creation of the WAC. One watershed resident characterized the Watershed Agricultural Program like this:

> It is my belief that the implementation of Whole Farm Planning will not only ensure cleaner water, but [also demonstrates] that a strategy of cooperation can accomplish more than heavy-handed regulations ... [17]

---

[14] Ad Hoc Task Force on Agriculture and New York City Watershed Regulations. 1991. *Policy Group of Recommendations*, December (unpublished manuscript), p. 1.

[15] Pfeffer, Max J., J. Mayone Stycos, Leland Glenna and Joyce Altobelli. 2001a. "Forging New Connections between Agriculture and the City." Pp. 419–446 in *Globalization and the Rural Environment*, edited by Otto T. Solbrig, Robert Paarlberg, and Francesco di Castri, Cambridge, MA: Harvard University Press;

[16] Pfeffer, Max J. and Linda P. Wagenet. 1999. "Planning for Environmental Responsibility and Equity: A Critical Appraisal of Rural/Urban Relations in the New York City Watershed." Pp. 179–206 in *Contested Countryside: The Rural Urban Fringe of North America*, edited by Mark B. Lapping and Owen Furuseth, Brookfield, VT: Ashgate.

[17] *Delaware County Times* (DCT) 3/15/94, pp. 1–2, 12.

A second key development was the formation of the Coalition of Watershed Towns (CWT). While challenges to NYC's proposed actions took many forms and involved a variety of actors, the CWT played a leading role in this opposition.[18] The CWT, organized in 1991, represented about 30 watershed towns west of the Hudson River. It was explicitly formed to protect west-of-Hudson watershed communities "from adverse impacts of the regulations, ensure that the City compensated watershed communities for direct and indirect costs of its watershed protection program, ensure that the regulations did not prevent reasonable community development, challenge the City to make the regulatory process fair, and limit the regulations to the minimum needed to protect water quality."[19] The CWT pursued legal action as one means of meeting its objectives.

The CWT helped articulate the common interests of watershed towns, but the development of a unified voice for communities was more difficult than it had been for the farm sector. The community effort started from a relatively fragmented local political structure based on home rule (i.e., the right to self-govern in local affairs), and the *town governments had no equivalent bridging institutions like the agricultural support agencies mentioned above*. Despite these handicaps, a nucleus for development of the CWT emerged in Delaware County.

The fact that the CWT emerged in Delaware County is no coincidence. Counties in New York have different political structures. Delaware County has a Board of Supervisors that administers the county, unlike the other watershed counties (e.g., Greene, Ulster, and Sullivan) that have directly elected County Legislatures. *The significance of this difference is that a board of supervisors served as a bridging institution between rural communities that may have few other formal ties*. The Board of Supervisors facilitated direct face-to-face contact between town supervisors and allowed them to share their concerns about NYC's proposed regulatory actions. In sharing these concerns, they identified common interests and their shared need for concerted action. Delaware County was thus an ideal base for the emergence of the CWT, which then went on to incorporate towns from throughout the watershed.

The importance of the CWT cannot be overemphasized. One Catskill resident remarked perceptively, *The regulations did us a favor. Their high-handedness and heavy restrictions brought the Coalition of Watershed Towns into being. The towns working together like that is remarkable, unheard of.* According to Stave[20]:

> Some felt the Coalition was possible because the regulations created a general feeling of social cohesion in the region; others felt the Coalition was driving that cohesion. Either way, Catskill residents have strengthened their bonds by asserting themselves against a perceived common threat.

---

[18]Finnegan, Michael C. 1997. New York City's Watershed Agreement: A Lesson in Sharing Responsibility. *Pace Environmental Law Review* 14:577–644.

[19]Stave, Krystyna Anne. 1998. *Water, Land, and People: The Social Ecology of Conflict over New York City's Watershed Protection Efforts in the Catskill Mountain Region*. PhD Dissertation, NY: Yale University, p. 179

[20]Stave, Krystyna Anne. 1998. *Water, Land, and People: The Social Ecology of Conflict over New York City's Watershed Protection Efforts in the Catskill Mountain Region*. PhD Dissertation, NY: Yale University, p. 295.

Tensions peaked when the CWT filed suit to prevent NYC from implementing its filtration avoidance plans. The CWT cited economic burdens on watershed residents resulting from restrictions placed on the use of privately owned land. It claimed that NYC would benefit almost exclusively from environmental measures in the countryside to protect drinking water supplies. In the face of this stalemate, EPA and other interested parties urged New York Governor Pataki to intervene to bring the interested parties to the negotiating table. These negotiations culminated in the NYC Watershed Memorandum of Agreement signed in January 1997.[21]

Institutionalization of a Watershed Community of Interests

The MOA codified a community of interest in several ways. It established a Land Acquisition Program that allowed NYC to purchase land or conservation easements, and in return NYC promised not to exercise eminent domain to acquire land for watershed protection. The Revised Watershed Rules and Regulations went into effect, and the agreement established the Watershed Protection and Partnership Programs as mechanisms for generating input about management of the NYC Watershed from a wide range of interests, all of which "share the common goal of protecting and enhancing the environmental integrity of the Watershed and the social and economic vitality of the Watershed communities."[22] These programs are the organizational representation of the watershed community of interests.

Prior to the signing of the MOA, there were differences of opinion about who legitimately had a stake in the management of the watershed and who had responsibility for its care. There were also disagreements about trends in environmental quality and the need to raise regulatory requirements. By the time the MOA was signed, the media began to report common interests between NYC and the watershed communities, and to express support for a partnership.

In the NYC watershed, the question of boundary conditions largely played out with respect to land; whose land would be subject to specific regulations or whose land NYC might purchase or condemn through eminent domain.[23] Prior to the MOA, the right of access to land was a contentious issue between the upstate towns and NYC. That sentiment began to change as the MOA was negotiated and public sentiment softened, as indicated by statements in local newspaper editorials:

---

[21]National Research Council (NRC). 2000. *Watershed Management for Potable Water Supply: Assessing the New York City Strategy.* Washington, DC: National Academy Press.

[22]New York City Watershed Memorandum of Agreement (MOA). 1997. January (unpublished manuscript), Article IV, Para. 97, p. 35.

[23]Pfeffer, Max J., Linda P. Wagenet, John Sydenstricker-Neto, and Catherine Meola. 2005. "Reconciling Different Land Use Value Spheres: An Example at the Rural/Urban Interface." Pp. 186–201 in *Land Use Problems and Conflicts: Causes, Consequences and Solutions* edited by Stephan Goetz, James Shortle and James Bergstrom, London: Routledge.

"...the deal includes ironclad assurances that the city will not condemn property, and will pay for the land it buys in a timely fashion. It's not gonna take 30 years for you to get your money like it used to."[24]

"The agreement eliminates the threat of big government staring over Catskill Mountains and East of Hudson River resident's shoulders."[25]

Clearly, the controversy over NYC's watershed management plan emerged when upstate communities claimed that NYC had historically treated them unfairly. Early in the controversy, sentiments expressed in the media indicated that the communities were being forced to bear a disproportionate share of the costs associated with water quality protection. For example, prior to negotiation of the MOA, one newspaper editorial declared: "There will be a price to pay and the people of the Catskills can't be asked to pay the price, and, second, our quality of life must be preserved with some kind of home rule procedure to protect it. No one will get what they want by making the Catskills a rural slum. People resenting their predicament won't work to protect the environment like an informed and empowered citizenry would."[26]

This sentiment changed in the process of negotiating the MOA. The MOA called for NYC to provide $240 million to the Catskill Watershed Corporation for infrastructure, economic development, conservation, education, and operations. With the adoption of the MOA, the media began to report the relationship with NYC as one providing resources, and that the responsibilities for protecting water quality would be shared more fairly. For example: "We're getting a lot out of this deal. We need to give something back in return."[27] "The MOA is the basis for the strong partnership between the parties, which ensures that watershed protection can be a win–win situation for all involved."[28] "Watershed Protection Planning promotes environmentally responsible future community planning which will minimize the adverse impacts on water quality and protect the economic vitality of the Watershed communities."[29]

Conclusion

The negotiation of a New York City (NYC) watershed management plan in the 1990s represented the creation of a watershed community of interest. This achievement demanded that the socially and politically isolated watershed communities come together to pursue common interests. There were some important obstacles

---

[24] *Catskill Mountain News* (CMN), 10/02/96, p. 16.

[25] *Delaware County Times* (DCT), 11/17/95, p. 2.

[26] *Catskill Mountain News* (CMN), 4/11/94, p 1.

[27] *Delaware County Times* (DCT), 4/01/96, p. 1.

[28] *Delaware County Times* (DCT), 8/04/00, p. 1.

[29] *Delaware County Times* (DCT), 11/10/95, pp. 1–17.

to such unification. The sparsely settled area had few resources and there was relatively little opportunity or perceived need for cooperation between communities. NYC's proposal to update the watershed rules and regulations presented a common and formidable threat demanding cooperation.

Preexisting social organizations played a pivotal role in allowing the communities to come together as a unified front. First, the Soil and Water Conservation District in heavily agricultural Delaware County and the State Department of Agriculture and Markets provided social resources to link farmers and leaders across political boundaries. The Watershed Agricultural Council (WAC) became the first upstate organization to negotiate with NYC over the terms and conditions of watershed management. Secondly, Delaware County was the nucleus of the Coalition of Watershed Towns (CWT). Formation of the CWT was facilitated by the existence of a County Board made up of supervisors from each town in the county. This form of government created the opportunity for political representatives to interact directly and to begin to articulate a shared set of grievances and demands. The CWT grew to represent towns throughout the major west-of-Hudson portion of the watershed in negotiations with NYC. Organizations like the WAC and CWT played crucial parts in the negotiations of the NYC Watershed Memorandum of Agreement, and the WAC was able to negotiate special terms for the regulation of agricultural activities in the watershed. These organizations contributed to the formation of a watershed community of interest through participation in the negotiation of common terms of reference, boundary conditions, and a fair solution. The result of this process was the creation of a watershed community of interest that recognized shared responsibilities and benefits associated with watershed management.

The important take-away lesson from this case study is that effective watershed protection requires collaboration among the communities and organizations that are part of the watershed. They must develop a watershed community of interest that they recognize and embrace as their own. No two watersheds are alike, so there are no simple recipes for creating a community of interest. Nevertheless, by mustering existing organizational resources that already work across local boundaries, watershed managers can address three issues essential for creating a community of interest: creating common terms of reference, agreeing on physical and legal boundaries defining natural resource use, and establishing standards of fairness and equity in implementing watershed protection measures.

Chapter 10
Upstream, Downstream: Forging a Rural–Urban Partnership for Shared Water Governance in Central Kansas

Theresa Selfa and Terrie A. Becerra

Introduction

Participatory governance of natural resources, including water, emerged in the 1980s under such names as collaborative watershed management, ecosystem management, grass-roots ecosystem management (GREM), watershed partnerships, and community-based natural resource management.[1] Watershed partnerships draw together diverse stakeholder groups to collectively negotiate the management of their water resources in a more proactive and egalitarian manner that is an alternative to traditional agency-driven water resource management and planning.[2] Although stakeholders, individually or as groups, may pursue self-interests, they also pursue the collective

---

[1] Weber, Edward P. 2000. "A New Vanguard for the Environment: Grass-Roots Ecosystem Management as a New Environmental Movement." *Society and Natural Resources* 13:237–259; Wollondeck, Julia M. and Steven L. Yaffee. 2000. "Making Collaboration Work: Lessons From a Comprehensive Assessment of Over 200 Wide Ranging Cases of Collaboration in Environmental Management." *Conservation in Practice* 1(1):17–25; Brick, Philip, Donald Snow, and Sarah Van De Wetering (eds). 2001. *Across the Great Divide: Explorations in Collaborative Conservation and the American West*. Washington, DC: Island Press.

[2] Leach, William D. and Neil W. Pelkey. 2001. "Making Watershed Partnerships Work: A Review of the Empirical Literature." *Journal of Water Resources Planning and Management* 127(6):378–385; Parisi, Domenico, Michale Taquino, Steven M. Grice, and Duane A. Gill. 2004. "Civic Responsibility and the Environment: Linking Local Conditions to Community Environmental Activeness." *Society and Natural Resources* 17(2):97–112; Sabatier, Paul A., Will Focht, Mark Lubell, Zev Trachtenberg, Arnold Vedlitz, and Marty Matlock (eds). 2005. *Swimming Upstream: Collaborative Approaches to Watershed Management*. Cambridge, MA: The MIT Press; Morton, Lois W. 2008. "The Role of Civic Structure in Achieving Performance-Based Watershed Management." *Society and Natural Resources* 21:751–766.

T. Selfa (✉)
Department of Environmental Studies, SUNY College of Environmental Science and Forestry, 107 Marshall Hall, Syracuse, NY 13210-2787
e-mail: tselfa@esf.edu

L.W. Morton and S.S. Brown (eds.), *Pathways for Getting to Better Water Quality: The Citizen Effect*, DOI 10.1007/978-1-4419-7282-8_10, © Springer Science+Business Media, LLC 2011

best interests of their community through these partnerships.[3] Through participation, citizens can be empowered to manage resources more consistently with their own objectives.[4] The rationale for a process of citizen participation is the belief that local citizens contribute knowledge, experience, and insight to local issues that lead to solutions more preferred by the public. In addition, citizens can gain a more sophisticated level of technical and social understanding, and agency administrators can learn which policies are acceptable to different community groups.

While public participation in water management decision making may be desirable in terms of better policy decisions and willingness to implement solutions, barriers to citizen participation may include complacency about the environmental issue and concerns about the lack of any real authority or decision-making power for citizens in the process.[5] In this context, institutional structures and their governance patterns affect the extent to which citizens perceive they can influence decisions as well as their willingness to become involved.

We examine a rural–urban watershed governance partnership, drawing on a policy network framework for our analysis.[6] We describe the origins, structure, and operations of the Citizens Management Committee (CMC) in the Cheney Lake watershed in south–central Kansas and its 15-year partnership with the city of Wichita to explain its institutionalization and success. The historical development of Cheney Reservoir, the conditions that led to the formation of Cheney Lake Watershed Inc. (CLWI), and its partnership with the city of Wichita provide the context for watershed management decision making and implementation of jointly developed plans.

Watersheds as a Management Unit

Integrated water resource management within the ecological boundaries of a watershed is one of the alternative management processes that emerged from the decentralization and participatory trends in water resource governance.[6] Barham[7] argues that using a geographical unit such as a watershed represents a change in perspective that ties the

[3]Parisi, Domenico, Michale Taquino, Steven M. Grice, and Duane A. Gill. 2004. "Civic Responsibility and the Environment: Linking Local Conditions to Community Environmental Activeness." *Society and Natural Resources* 17(2):97–112.

[4]Spaling, Harry. 2003. "Innovation in Environmental Assessment of Community-Based Projects in Sub-Saharan Africa." *Canadian Geographer* 47(2):151–168; Parisi, Domenico, Michale Taquino, Steven M. Grice, and Duane A. Gill. 2004. "Civic Responsibility and the Environment: Linking Local Conditions to Community Environmental Activeness." *Society and Natural Resources* 17(2):97–112.

[5]Irvin, Renee A. and John Stansbury. 2004. "Citizen Participation in Decision Making: Is It Worth the Effort?" *Public Administration Review* 64(1):55–65.

[6]Ferreyra, Cecilia, Rob C. de Loe, and Reid D. Kreutzwiser. 2008. "Imagined Communities, Contested Watersheds: Challenges to Integrated Water Resources Management in Agricultural Areas." *Journal of Rural Studies* 24:304–321.

[7]Barham, Elizabeth. 2001. "Ecological Boundaries as Community Boundaries: The Politics of Watersheds." *Society and Natural Resources* 14:181–191.

human community to the natural community, and that it has been shown to encourage a sense of environmental responsibility. Traditional social systems may serve as constraints to watershed planning if conservation or management goals are not working toward the same ends. However, the transition to a watershed approach must be accompanied by the creation of more democratic institutions and processes to avoid creating social injustices. Such a shift in human understanding of natural systems and our place within them suggests that environmental problems require social and political solutions as much as technical solutions.

Nonetheless, using watershed boundaries for environmental planning and policy implementation is not without challenges because ecological and political boundaries rarely coincide.[6] Difficulties with using the watershed as a management planning unit include the potential for political disputes when sociopolitical units overlap,[8] a perceived threat to property rights, and competition between municipalities for economic development projects. Using the watershed as a planning unit also may raise questions about who is included or excluded, what the structure of participation should be, the mechanism by which policymakers will be held accountable, and which local agencies are responsible. Ferreyra et al.[6] also caution that the imposition of a watershed-scale approach driven by ecological imperatives may actually reduce diversity of interests by sidelining important actors who are not part of the "community of place" but are part of the "community of interest" in the watershed.

Morton[9] contends that incentive-driven and regulatory measures can also undermine democratic processes of watershed management by taking away social ownership and responsibility from the local residents. She suggests that a model is needed in which individuals in a community collectively share ownership of the problems and the responsibility for finding the solution. More directly, Conca,[10] in discussing governance of water resources globally, posits that when established institutional forms are inadequate, new institutional forms may arise from grassroots networks, coalitions, citizens' organizations, activists groups, or social movements that may create mechanisms for environmental governance outside established institutions. Our case study of the formation and evolution of CLWI and the CMC is an example of the emergence of such a new governance structure.

A Policy Network Approach to Water Governance

Policy networks as described by Ferreyra et al.[6] represent a specific form of watershed governance that implement the policy-making process through interactions

[8]Blomquist, William. and Edella Schlager. 2005. "Political Pitfalls of Integrated Watershed Management." *Society and Natural Resources* 18:101–117; O'Neill, Karen. 2005. "Can Watershed Management Unite Town and Country?" *Society and Natural Resources* 18(3):241–253.

[9]Morton, Lois W. 2003. "Civic Watershed Communities: Walking Toward Justice: Democratization in Rural Life." *Research in Rural Sociology and Development* 9:121–134.

[10]Conca, Ken. 2006. *Governing Water: Contentious Transnational Politics and Global Institution Building*. Cambridge, MA: The MIT Press.

and reciprocal exchanges. In a policy network, policies are made and implemented based on a horizontal network of interests in nonhierarchical relationships. Decision making within these networks is achieved by negotiation and consensus rather than the conventional top–down method of policy making. Instead, policy networks focus on the interactions of individuals, institutions, and ideas within the same and among actors within different policy sectors that share a common standpoint.

A policy network approach describes the policy-making process across multiple levels of actor networks. The "rules" of a policy network set the guidelines for stakeholders' participation in the decision-making process. For example, county-level agencies, producers, and cities have varying resources, but these differences in resources do not necessarily give one stakeholder more power than another within the network. Agencies may take a leadership role, but in a successful policy network are no more powerful in the decision-making process than producers or other stakeholders. The equality of status rule among stakeholders creates the basis for nonhierarchical relationships among the network actors.[11]

In Cheney Lake Watershed, policy sectors including conservation, agriculture, and urban water supply are represented in the network, all of whom have similar degrees of status and power in the network. The actors within the network reached consensus or "shared frame of reference"[12] that implementing best management practices (BMPs) is an effective management approach to protect soil quality at the field level and to protect water quality at local, regional, and national levels.

Methods

Historical background on the origins of Cheney Lake Watershed was derived from documents detailing the agricultural and land practices used in the watershed throughout the past 15 years and through interviews with key stakeholders. Historical documents regarding the establishment and the current and past missions of the Kansas Ground Water Management Districts, the history of activities of the Natural Resource Conservation Service in Kansas, US Bureau of Reclamation accounts of the Cheney Project, and anecdotal and written accounts of the circumstances surrounding the construction of Cheney Reservoir were also analyzed.

---

[11]Ferreyra, Cecilia, Rob C. de Loe, and Reid D. Kreutzwiser. 2008. "Imagined Communities, Contested Watersheds: Challenges to Integrated Water Resources Management in Agricultural Areas." *Journal of Rural Studies* 24:304–321; Sabatier, Paul A., Will Focht, Mark Lubell, Zev Trachtenberg, Arnold Vedlitz, and Marty Matlock (eds). 2005. *Swimming Upstream: Collaborative Approaches to Watershed Management.* Cambridge, MA: The MIT Press.

[12]Ferreyra, Cecilia, Rob C. de Loe, and Reid D. Kreutzwiser. 2008. "Imagined Communities, Contested Watersheds: Challenges to Integrated Water Resources Management in Agricultural Areas." *Journal of Rural Studies* 24:304–321, p. 305.

Additional data on the history of water governance in Kansas was gathered through in-depth interviews with 6 current and past members of the Cheney Lake Citizens' Management Committee and with 26 agricultural producers in the watershed. Four additional interviews were held with Kansas Water Office and Natural Resources Conservation Service (NRCS) personnel. The interviews with state officials were designed to elicit information about the history of water governance in the State of Kansas, while the interviews with CMC members focused on how the organization started, their motivations for participating, and what CMC members see as successes and challenges faced by the CLWI. As part of a larger USDA Conservation Effects Assessment Program (CEAP) research project, we interviewed farmers about their perspectives on how effective CLWI has been in informing them about water issues and in facilitating positive behavior change.[13]

The Origin of Cheney Lake and Cheney Lake Watershed Inc.

Cheney Lake Watershed encompasses 633,000 acres and extends across five counties in south central Kansas (Fig. 10.1). More than 99% of the land in Cheney Lake Watershed is used for agricultural purposes, including both cropland and livestock pasture. Farmers and ranchers use 52% of the land area to produce corn, grain sorghum, soybeans, and wheat as well as livestock. The current human population within the watershed is less than 4,000,[14] and towns within the watershed range in size from 200 to slightly more than 1,200. Because of the small human population, the potential for point source pollution is considered small, although approximately 1,000 farms located in the watershed include small dairy and cattle feeding operations.[15]

Cheney Reservoir is a major public water supply for Wichita, the largest city in the state. Water supply has been a longstanding problem for Wichita since the early

---

[13] The interview questions were pretested in interviews with three current CMC board members. The questions were appropriately modified before conducting further interviews. CLWI staff provided 63 names of producers within the watershed. The list was sorted by sub-watersheds and we selected 32 names that were dispersed throughout the watershed (Fig. 10.1) and that captured the diversity of scale and types of operations existing in the watershed. Different cultural groups within the watershed were intentionally included; Potential subjects were mailed a letter describing the research and notified that a researcher would be contacting them by telephone to request an interview. Interviews were conducted in five sets between February 19 and March 14, 2008. The interviews were audio recorded for later transcription. Transcripts of the interviews were coded using QSR NVivo 7 software. The data was coded into themes within topical categories about the CMC, cost share, watershed knowledge, information sources, and farm management. Patterns, similarities, and differences were identified in how subjects talked about a particular topic and differences in values associated with those subjects.

[14] Devlin, Daniel, Nathan Nelson, Lisa French, Howard Miller, Philip Barnes, and Lyle Frees. 2008. "Conservation Practice Implementation History and Trends." Kansas State University Agricultural Experiment Station and Cooperative Extension Service, Publication # EP157. Manhattan, KS (http://www.oznet.ksu.edu/library/h20ql2/ep157.pdf).

[15] There is one permitted cattle feeding operation in the watershed located that is located in Turon.

1920s, when a study showed that water from the Arkansas River, then the city's water supply source, contained such high concentrations of minerals that continued use of river water would cause Wichita to incur increased maintenance costs for the water system.[16] As an alternative, it was suggested Wichita draw its water from the Equus Beds, a local aquifer in south–central Kansas that extends northeast of the watershed through Reno, Sedgwick, Harvey, Marion, and Rice Counties, and is primarily recharged by annual precipitation and seepage from overlying river and streambeds.

Population and industrial growth in Wichita from 1938 to 1946 necessitated an increased water supply, because the Equus Beds, the city's primary supply source by 1940, was no longer adequate. By 1950, the city was again looking for a new supply source. A drought that same year caused farmers to drill irrigation wells to get water for their crops. The situation worsened in 1954 when the city experienced another water emergency.[17]

Meanwhile in 1947 the United States Bureau of Reclamation had begun a basin-wide study of the Arkansas River to identify possible reservoir sites that could provide water for irrigation, municipal, and industrial uses. In 1952, the Bureau

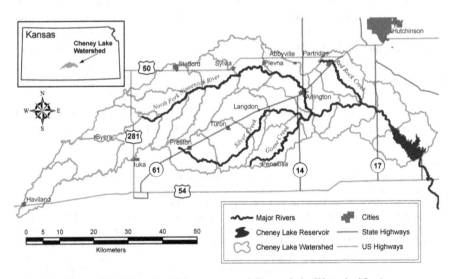

Map 10.1   Cheney Lake Watershed (Map courtesy of Cheney Lake Watershed Inc.)

---

[16]Gattin, Timothy D. 1995. "Cheney Reservoir: Creating and Preserving a Surface Water Supply for the City of Wichita." Unpublished Masters' Thesis. Hugo Wall Center for Urban Studies, Wichita State University, Wichita, KS.

[17]Hufford, D. Brian. 1982. *The Ayes Have It, Wichita Water Department, A History 1882–1982.* Wichita, KS: Frank Wright-Josten.

recommended that the Ninnescah River in Reno County, Kansas be damned on the North Fork at Cheney, to create Cheney Reservoir. Later that same year, the Wichita City Commission asked the Bureau to investigate the economic and engineering feasibility of a water supply from the Ninnescah River.

In 1956, the citizens of Wichita voted in favor of revenue bonds to support the Cheney project. Presidential approval for legislation authorizing the reservoir was signed in 1960 and construction was completed in the spring of 1965.[17] The reservoir was designed as a 100-year multipurpose project. In addition to being a water supply, it provides recreational opportunities, wildlife habitat, and agricultural benefits through flood control to several hundred thousand Kansas residents. The incorporated flood control measures provide for 3,700 acres of farmland to be irrigated, although no stored water was intended for irrigation.[18] The city of Wichita currently draws more than 60% of its daily water supply from the reservoir.[19] The water supply is also marketed to other smaller cities in the Wichita area.

Cheney Lake Watershed has been the focus of organized water quality protection efforts for longer than 15 years. In the early 1990s, the city of Wichita became concerned about the amount of phosphorus in Cheney Reservoir. The high levels of phosphorus were contributing to algae blooms in the lake, which in turn were creating taste and odor problems in city water. Water treatment personnel determined that the phosphorus from soil was coming into the lake with sediment. At the same time, members of the Reno and Sedgwick County Conservation District boards were concerned about soil loss in the watershed, as they had observed sediment deposition in Cheney Reservoir and stream bank erosion throughout the watershed. Understanding the significance of the lake as Wichita's water supply and the distress of displacing families when the reservoir was built, members of the conservation district boards (many of whom are also local producers) were concerned that the city of Wichita would condemn more land in the watershed in efforts to reduce sediment and phosphorus in the lake.

When the city of Wichita and the conservation districts boards both recognized that these water issues within the watershed needed to be addressed, the stage was set for a rural–urban partnership. Agency staff, farmers, ranchers, and other rural residents within the Cheney Lake watershed came together in 1999 to create CLWI, a private, nonprofit organization that works cooperatively with the city of Wichita to protect the water quality of Cheney Reservoir. The organization promotes farm management practices that conserve soil and protect surface and ground water from farm chemicals and sediment. The organization is governed by a seven-member board, the Cheney Lake Watershed Citizens' Management Committee, each board member is a producer elected from one of the seven sub-watersheds.

---

[18] Bureau of Reclamation, 2008 (http://www.usbr.gov/dataweb/html/Wichita.html#general).

[19] City of Wichita. 2009. Cheney Reservoir Environmental Assessment Summary (http://www.wichita.gov/CityOffices/WaterAndSewer/ProductionAndPumping/Cheney.htm).

Water Governance: Forging an Urban–Rural Partnership

The CLWI is a product of cooperation between rural and urban communities that originated with concerns about preserving agriculture and community from one side and the need for a growing city to solve an urban water quality problem on the other. While the relationship between the CLWI and the city of Wichita may seem to be somewhat unusual, it builds on Kansas' uncommon approach to water governance. Kansas differs from other states in that its primary water regulatory authority remains within the Kansas Department of Agriculture. While such authority may be unusual, irrigation still accounted for 83.8% of the Kansas' total water use in 2000.[20] Historically Kansas has diffused the responsibilities for water governance, distributing them through several state-level offices, local-level groundwater management districts, and, to a lesser degree, irrigation districts, and rural water districts. The rationale is that diffusing authority provides more checks and balances to water governance.[21]

The policy network framework[6] can be used to examine the evolution of the governance structure of the Cheney Lake watershed within multiple levels of actor networks. Water management personnel of the city of Wichita work with the city government staff and with the CMC, which draws its members from producers in the watershed to ensure that all sub-watershed areas are represented. The CMC works with other producers as well as with personnel from agencies like the NRCS. The CMC and staff members of CLWI facilitate changes in farm management practices by mediating between agencies and farmers. CLWI has been able to provide feedback on farmer needs when rigid program requirements inhibit producer adoption. This has resulted in increasing program flexibility.

The CLWI emerged from a task force that was developed in 1992 by the conservation districts and the city of Wichita to identify and alleviate potential sources of pollution in the watershed and Cheney Reservoir. The initial task force consisted of Wichita city government; Wichita Water and Sewer Department; federal, state, and county agencies; and a committee of landowners.[22] Initially, some conservation district board members and some landowners held suspicions about Wichita's motives in agreeing to the partnership. To minimize or avert potential conflict between the two sides, the conservation district members decided to establish a separate subcommittee to guide the watershed project. The subcommittee was largely composed of the landowners that served on the initial task force, which became the Citizen's Management Committee; the CLWI grew from there.

The city and producers in the watershed reaching a common understanding of the problem or a "shared frame of reference" – deteriorating water quality in Cheney Reservoir as a product of sediment runoff – influenced the perceptions of

---

[20] Kenny, Joan F. and Cristi V. Hansen. 2004. "Water Use in Kansas, 1990–2000." U.S. Geological Survey (http://pubs.water.usgs.gov/fs20043133).

[21] Kansas Water Office, 2008.

[22] Cheney Lake Watershed Inc. (CLWI). 2006. North Fork Ninnescah River Project Management Plan.

the actors, helping them to define the problems they needed to address, what management strategies to use, and which stakeholders to include in forging the solution. The city's initial plan to reduce sediment and phosphorus was to construct sediment retention dams throughout the watershed; prevention was preferable to remediation. The conservation district board members offered as an alternative solution that the conservation district personnel approach producers with a plan to stop the erosion at the field. The city of Wichita agreed to the suggestion and the conservation districts put together a plan.[23]

This mutual interest between Wichita and residents of Cheney Lake Watershed regarding the water quality of Cheney Lake has forged a unique arrangement that allows the agricultural-centered leadership of the watershed – the CMC – to direct the water protection efforts without direction or oversight by the city of Wichita, despite the city's greater financial contribution.[23] Further, recognizing that those conservation efforts will increase production costs for producers, but that those practices are in the city's best interests, the city contributes substantial financial support to producers to help abate the costs of adopting water quality protection practices.

As actors in the policy network, agencies contribute their expertise and cost share programs as tools that CLWI can draw on in their efforts to promote BMPs that reduce soil erosion (sediment and chemical runoff) and water contamination from livestock waste. Between 1994 and 2006, a cumulative 1,369 BMP contracts have been adopted on 700 sites,[14] and research is currently underway to determine the impact of the conservation efforts initiated by the partnership between CLWI and the city of Wichita.[24] Similarly, landowners and producers in the watershed contribute local knowledge and leadership to efforts to educate other landowners and producers about the water quality issues of Cheney Lake, the benefits they can gain in their operations by adopting BMPs, and the mutual benefits of voluntarily protecting water quality.

The CLWI members attribute the success of the organization in gaining buy-in from producers and from the city to its emphasis on local control and reliance on members' local knowledge. The CLWI is also attributed with facilitating more flexibility with requirements for cost-shared programs. According to one producer: "This office (CLWI), through the citizen management committee, has been pretty good about adjusting those management plans – One to be a little more farmer friendly and still control water quality – the flexibility has got to come in the management plan."[25]

Reliance on local knowledge is also mentioned by several producers as integral to the success of the organization in the watershed. For example, "I know my neighbor's land. And with our board, until you get way, way out west. I really doubt if there's many fields in that whole watershed that we can't tell you what's on that field. Or where that water goes, or which way it goes and where it washes."[25]

---

[23] Interview 702081, 2008.

[24] Nelson, Nathan, Kyle Mankin, Michael Langemeier, Daniel Devlin, Philip Barnes, Theresa Selfa and William Hargrove. 2006. "Assessing the Impact of a Strategic Approach to Implementation of Conservation Practices." USDA-CSREES Conservation Effects Assessment Program, 2006–2009.

[25] Interview 100083, 2008.

A Mutually Beneficial Rural–Urban Partnership

The farmers in the Cheney Lake watershed who are members of CLWI see the efforts of the organization as not only benefitting themselves and the city of Wichita, but also as both sides taking ownership and responsibility for their own well-being. The city recognizes that Cheney Reservoir water is a public good for which the farming community cannot bear sole responsibility.[26] The farmers understand that adopting BMPs is good stewardship that is in their own best interest as well as in the interests of the citizens of Wichita. They see voluntary adoption of the management practices as a way to prevent further regulation of their practices and, ultimately, their livelihoods. It also reinforces the importance of local control to many of the farmers and informants we interviewed. When asked about the responsibility of keeping rivers and streams throughout the watershed and the reservoir clean, one producer expressed the broadly held sentiment: "I'd rather be trying to get it done voluntarily than have those 500,000 people come up here and tell us what we're going to do."[25] In addition, an extensive, long-term education component goes along with persuading farmers to change their management practices and contributes to behavior change – and the attitudes, beliefs, and perceptions that go along with behavior change. According to a producer that has been active in the organization, "We have producer meetings, and we have seminars, we have field days…. In a year's time, I would hate to guess how many hours I spend sitting along the side of the road talking to a producer about his situation. Just a lot of word-of-mouth."[25]

The members of CLWI have formed a watershed-focused community of interest that extends across five counties of the watershed and includes state and federal agency staff, producers, and other watershed residents. They have created a new institution in CLWI that is not restricted by county boundaries. Decisions result from equitable input from watershed stakeholders: the representatives of the various local, state, and federal agencies; the producers, landowners, and other residents of the watershed; and representatives of the city of Wichita. The city of Wichita and agency stakeholders may have more financial resources and/or expertise to contribute to the watershed efforts, however, they are no more influential or powerful in the decision-making process than are landowners, producers, or watershed residents, as was also found in other related research.[11] "It was probably the NRCS office and the local conservation people that saw the need and wanted to see if Wichita would help fund some of the changes that needed to happen to improve their water quality. But they had an idea, and they brought people together and began to say, 'It has to be an individual, producer driven; it can't be an agency driven thing'."[27]

While the CMC and CLWI promote adoption of BMPs and farm programs to help producers with the cost, each producer makes decisions situationally.

---

[26] http://cheneylakewatershed.org/ruralurban, 2006.

[27] Interview 100082, 2008.

One producer offers this explanation of how CLWI works, "The CMC is involved, but they don't make our decisions; they are an educational tool. The NRCS office, Farm Service Agency, and extension office would have some authority advisory and so forth. The Kansas Department of Wildlife and Parks would probably consider they had valuable input to put into conservation things as it relates to wildlife."[27] Producer perspectives are incorporated into the decision-making process through the CMC board members. Several producers expressed how the CMC represents the farmers in the watershed, "With the help of the CMC, we do have a voice for local farmers and businessmen and we can voice our opinion on water quality and implementation of practices, whether they are feasible or not."[28] And, "When I was on the CMC, that was one of the things I pushed the city of Wichita to allow, the flexibility to work with individual producers."[29] This equality of status among stakeholders creates the basis for nonhierarchical relationships among the network actors.

Conclusion

The Cheney Lake Watershed case study illustrates the mutual benefits that can be derived from different stakeholders working together for a common goal from both inside and beyond the watershed. The policy network approach allows us to move beyond the geographic and ecological boundaries of the watershed to examine the networks in communities of place and communities of interest, and how the water resource concerns of the city of Wichita and the concerns of the agricultural producers in the larger Cheney Lake Watershed can be integrated. When the needs and capacities of the different networks addressing water resources throughout the entire watershed were incorporated into policy making, common ground was found and a more satisfactory, longer-term solution was implemented. Out of the process, the CLWI emerged as a new institution for watershed governance. The partnership between the agricultural producers of Cheney Lake Watershed and the city of Wichita has resulted in both sides recognizing that they share ownership and responsibility for the long-term sustainability of Cheney Lake Reservoir. When viewed through the policy network approach, this case study demonstrates the benefits of not presupposing the scale at which such partnerships work, or at which natural resource governance occurs, but rather looking at vertical and horizontal linkages that emerge across space.

Acknowledgments We would like to thank Lisa French and Howard Miller of the Cheney Lake Watershed Inc. and all of the farmers in the watershed who were willing to talk with us. The research was funded by USDA CEAP program.

---

[28] Interview 219081, 2008.

[29] Interview 303082, 2008.

Chapter 11
Local Champions Speak Out: Pennsylvania's Community Watershed Organizations

Kathryn Brasier, Brian Lee, Richard Stedman, and Jason Weigle

Introduction

Community-based resource management (CBRM) is a process based on local stakeholder involvement in the development of policies and programs and direct amelioration of environmental problems. Community watershed organizations (CWO) in Pennsylvania are grassroots driven and a form of CBRM. CBRM has been applied across all classes of natural resources, including forests,[1] wildlife, water resources, fisheries,[2] land, and agriculture.[3] These local efforts grew out of frustrations and conflicts over national-level conservation policies and regulations being imposed on local communities' use of natural resources.[4] Community-based conservation movements burgeoned in the 1990s as a result of several factors: increased awareness of environmental issues in the 1960s and 1970s, increased

---

[1] Pagdee, A., Y.S. Kim and P.J. Daugherty. 2006. "What makes community forestry management successful: a meta-study from community forests throughout the world." *Society & Natural Resources* 19:33–52.

[2] Satria, A., Y. Matsuda and M. Sano. 2006. "Questioning community based coral reef management systems: case study of *Awig-Awig* in Gili Indah, Indonesia." *Environ Dev Sustain* 8:99–118; Armitage, D.R. 2005. "Community-based narwhal management in Nunavut, Canada: change, uncertainty and adaptation." *Society & Natural Resources* 18:715–731.

[3] Kellert, S.R., J.N. Mehta, S.A. Ebbin and L.L. Lichtenfeld. 2000. "Community natural resource management: promise, rhetoric, and reality." *Society & Natural Resources* 13:705–715; Lubell, M. 2004. "Collaborative watershed management: a view from the grassroots." *Policy Stud J* 32:341–361.

[4] Lubell, M. 2004. "Collaborative watershed management: a view from the grassroots." *Policy Stud J* 32:341–361.

K. Brasier (✉)
Assistant Professor of Rural Sociology, Department of Agricultural Economics &
Rural Sociology, Penn State University, 105B Armsby Bldg, University Park, PA 16802, USA
e-mail: kbrasier@psu.edu

L.W. Morton and S.S. Brown (eds.), *Pathways for Getting to Better Water Quality: The Citizen Effect*, DOI 10.1007/978-1-4419-7282-8_11,
© Springer Science+Business Media, LLC 2011

activism by indigenous groups across the globe, and overall movement within governments to devolve conservation to the local level.[5]

Collaborative management approaches that include local citizens are meant to provide an alternative to "centralized, command-and-control policies"[6] that are more flexible, responsive, efficient, effective, and inclusive. CBRM provides a mechanism to coordinate multiple organizations, agencies with overlapping jurisdictions, and resource users. It also can fulfill federal requirements for public input in natural resource decision making (such as the National Environmental Protection Act). By including stakeholders throughout the process, partnerships can: (1) increase the flow of information between resource users and local, state, regional, and federal policy makers; (2) allow for effective conflict resolution; and (3) create more technically sound and locally acceptable solutions.[7]

A variety of CBRM efforts exist, from formal stakeholder partnerships and governance groups to grass-roots groups of private citizens. Stakeholder partnerships are coalitions of multiple organizations, user groups, agencies, and governmental units that meet routinely to discuss and negotiate management issues.[8] These groups have a formal, explicit position within the governance and management of the natural resource.[9]

Grass-roots watershed organizations are locally based groups of volunteers who are committed to a specific resource or vision for that resource.[10] Participating volunteers tend to be relatively homogeneous, and membership is usually composed of private citizens (but may also include representatives of related environmental or sportsmen's organizations, natural resource agencies, and municipalities). Watershed groups usually have no formal governance role. Instead, their goals tend to target changing the values, knowledge, and behaviors of land users.[11] Watershed groups

---

[5]McCarthy, J. 2005. "Devolution in the woods: community forestry as hybrid neoliberalism." *Environ Plann A* 37:995–1014; Western D. and R.M. Wright. 1994. "The background to community-based conservation." In: Western, D, and R. M. Wright (eds) (Strum SC (assoc. ed)) *Natural connections: perspectives in community based conservation.* Island Press, Washington, DC, pp. 1–12.

[6] Lubell, M. 2004. "Collaborative watershed management: a view from the grassroots." *Policy Stud J* 32:341–361.

[7]Bidwell, R.D. and C.M. Ryan. 2006. "Collaborative partnership design: the implications of organizational affiliation for watershed partnerships." *Soc Nat Res* 19:827–843; Leach, W.D. and N.W. Pelkey. 2001. "Making watershed partnerships work: a review of the empirical literature." *J Water Resour Plann Manag* 127(6):378–385.

[8]Leach, W.D. and N.W. Pelkey. 2001. "Making watershed partnerships work: a review of the empirical literature." *J Water Resour Plann Manag* 127(6):378–385; Leach, W.D., N.W. Pelkey and P.A. Sabatier. 2002. "Stakeholder partnerships as collaborative policymaking: evaluation criteria applied to watershed management in California and Washington." *J Policy Anal Manag* 21(4):645–670.

[9]Weber, E.P. 2003. *Bringing society back in: grassroots ecosystem management, accountability, and sustainable communities.* MIT, Cambridge, MA.

[10]Born, S.M. and K.D. Genskow. 2000. "The watershed approach: an empirical assessment of innovation in environmental management." *National Academy of Public Administration Research Paper Number 7.* http://www.napawash.org; Leach, W.D. and N.W. Pelkey. 2001. "Making watershed partnerships work: a review of the empirical literature." *J Water Resour Plann Manag* 127(6):378–385.

[11]Morton, L.W. 2003a. "Civic watershed communities: walking toward justice: democratization in rural life." *Res Rural Sociol Dev* 9:121–134.

can be seen as a new form of the environmental movement – one that is grass-roots and place based, and focused on the health of both the local environment and the community that depends on it.[12] These groups form because the founders see a vacuum of action or regulation to protect, restore, or preserve a particular resource.

We describe the outcomes of CBRM efforts on the management of rivers and streams and their surrounding watersheds in Pennsylvania. Although the ultimate goal of watershed groups is the improvement of the natural resource base, measuring and attributing positive environmental impacts can be difficult. There may be a significant time lag between restoration efforts and changes in water quality indicators. The groups' social impacts, such as changing behaviors that have long-lasting benefits for the resource, are valuable intermediary measures of success toward achieving water quality goals and may produce other unintended benefits as well.[13] Therefore, we focus on intended and unintended outcomes resulting in both environmental improvements and social or political changes.

After describing these outcomes, we then discuss the relationships watershed organizations have with local agencies and how these relationships in turn affect the groups' achievements.[14] This overview will provide useful guidance for professionals working with community watershed groups. In addition, personnel working in communities where watershed groups exist might consider ways of approaching these groups that can be most beneficial for the group, the agency, the community, and the natural resource base.

Studying Community Watershed Organizations in Pennsylvania

We define CWOs as nongovernmental, nonprofit, voluntary organizations with or without paid staff that work in a watershed at least partially in Pennsylvania, with water-related issues as a theme or mission.[15] As of 2002, 580 CWOs were identified in Pennsylvania.[15] The research reported here draws from in-depth

---

[12]Sirianni, C. and L.A. Friedland. 2005. *The civic renewal movement: community-building and democracy in the United States*. Charles F. Kettering Foundation, Dayton, OH; Weber, E.P. 2000. "A new vanguard for the environment: grass-roots ecosystem management as a new environmental movement." *Society & Natural Resources* 13:237–259.

[13]Stedman, R., B. Lee, K. Brasier, J. Weigle and F. Higdon. 2009. "Cleaning up water? Or building rural community? Community watershed organizations in Pennsylvania." *Rural Sociology* 74(2):178–200.

[14]Bidwell, R.D. and C.M. Ryan. 2006. "Collaborative partnership design: the implications of organizational affiliation for watershed partnerships." *Soc Nat Res* 19:827–843; Leach, W.D. and N.W. Pelkey. 2001. "Making watershed partnerships work: a review of the empirical literature." *J Water Resour Plann Manag* 127(6):378–385; Moore, E.A. and T.M. Koontz. 2003. "A typology of collaborative watershed groups: citizen-based, agency-based, and mixed partnerships." *Society & Natural Resources* 16(5):451–460.

[15] Lee, B. 2005. *Pennsylvania community watershed organizations: form and function*. Doctoral dissertation, The Pennsylvania State University.

interviews with leaders from 28 CWOs conducted in 2003 and 2004.[16] Groups
were selected to represent rural areas within all major state river basins (see Fig. 11.1).
All CWOs within each river basin were grouped into three categories (high, medium,
or low) based on the number of partners they reported during a 2002 mail survey. The
CWOs interviewed were randomly selected from within these categories.

Interviews with CWO leaders were digitally recorded and transcribed word-
for-word. The research team developed codes for the textual data in the transcripts
that reflected common themes within the interviews as well as related literature.
These codes were assigned to paragraphs within the transcripts. Summary documents of each code were prepared and discussed within the research team, to
further identify common threads or points of comparison. Direct quotations from
the interviews are in below.

Environmental and Community Outcomes

Positive outcomes from community watershed groups can be thought of as either
intended (fulfilling the stated goals of the group) or unintended.[17] The top reasons
cited for forming CWOs in Pennsylvania are improving ecosystem health
(especially water quality) and environmental education.[18]

Intended Outcomes: Improving Ecosystem Health

Measuring a group's direct impact on water quality is difficult, because there may
be a time lag between implementation and effect.[13] Further, some problems are
easier to treat than other problems. However, a few groups identified specific successes in changing water quality as measured through monitoring data. One CWO
leader says, "I think every project that they've undertaken in the way of abandoned
mine drainage … has been successful … you can … see the concentrations from
iron manganese to aluminum to pHs…. The data indicates both at the site and
downstream that the treatments are working."

For those not describing specific ecosystem changes from their activities, some
CWO leaders describe successful implementation of actions that are expected to

---

[16]The interviews reported here represent the third wave of research on CWOs in Pennsylvania.
Earlier waves included interviews with support organizations, funders, and CWO leaders, and a
mail survey sent to all 580 CWOs. For more details on the methods, see: Lee, 2005; Higdon,
Brasier, Stedman, Lee, and Sherman, 2005.

[17]Cable, S. and B. Degutis. 1997. "Movement outcomes and dimensions of social change: the
multiple effects of local mobilizations." *Curr Sociol* 45(3):121–135.

[18]Higdon, F., K. Brasier, R. Stedman, B. Lee and S. Sherman. 2005. *Assessment of community
watershed organizations in rural Pennsylvania.* Center for Rural Pennsylvania. http://www.
ruralpa.org/watersheds_higdon.pdf.

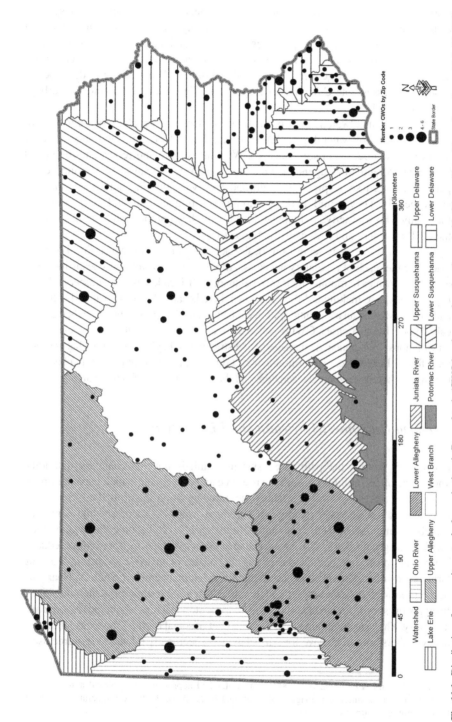

Fig. 11.1 Distribution of community watershed organizations in Pennsylvania (CWO locations are based on the centroid of the mailing address zip code as of July 23, 2002. Approximately 100 CWOs (25%) have post office boxes for which a geographic centroid could not be produced. Therefore not all CWOs are represented on the map)

lead to environmental improvements.[19] These projects include installation of systems to treat runoff from abandoned mines, cost-sharing, and volunteer installation of agricultural best management practices such as streambank fencing or riparian buffers, wetland construction, and streambank stabilization. One important activity for several groups is garbage removal from illegal dumps. When asked to describe the group's successes, one CWO leader responded, "Fifty thousand pounds of garbage out of the river!"

CWOs become "watchdogs," protecting a resource for which there are few advocates: "We kind of watchdog.... We really are kind of the guys and gals that look after that resource...." In some cases, CWOs investigate and report environmental violations to regulatory agencies. One leader interviewed says, "...a mining company had buried some pipes underground and was pumping acid water into the creek secretly. They would do this pumping operation at night, and ... an employee told a guy who was active in that watershed. And they got the Fish Commission and the DEP involved, and it ... was the biggest fine..."

CWOs in some cases serve as a locally acceptable place to report illegal activity, extending the reach of regulatory agencies. Additionally, their influence leads individuals and companies to monitor their own behavior: "DEP called us last summer and said, hey, you guys are making a difference down there because [a local business] called us and said ... we had an accidental spill, we'd better tell you before [the watershed group] called you." In other cases, CWOs police the area, and force regulatory agencies to act on violations: "We police the management district when they're not doing what we want to. If the ... Fish & Boat Commission aren't doing what we want to we put a big ad in the paper where we agitate the senators and the representatives from the area."

Intended Outcomes: Environmental Education

For several groups, the path to ecosystem health is through changing individual knowledge of and attitudes toward the resource and consequently their behavior on the landscape: "success ... isn't necessarily measured by an on-the-ground project ... you can measure maybe water quality has improved..., but ... success to me is making somebody more aware and ... giving them the tools ... to become that way on their own." Some groups worked to change resource users' practices by, for example, educating boaters on techniques for preventing the spread of invasive species or teaching farmers new management strategies to minimize sediment and nutrient loss. Nearly all of the groups interviewed report conducting educational events (presentations, hikes, fishing workshops, stream clean-ups, canoe trips, etc.) teaching about the resource and its history. They are attempting to foster a new set of social norms – the expectations we believe others around us have for our behavior – specifically related to resource use.

---

[19]Born, S.M. and K.D. Genskow. 2000. "The watershed approach: an empirical assessment of innovation in environmental management." National Academy of Public Administration Research Paper Number 7. http://www.napawash.org.

Many groups prioritize changing the behavior of future resource users – children. "The future belongs to our children and you've got to educate the children...." CWOs work with educators to garner resources to enhance environmental education offerings in schools. School districts, scouting groups, and other children's groups work directly with CWOs to install monitoring equipment and collect data, among many projects. These projects have profound impacts on participants and watershed group members: "...at least two of the three kids that are going have expressed interest in this type of work as a career, or at least a course of study.... I mean, that's the whole reason we do it. Period. It's to generate the interest in the young people."

In addition to direct impacts on individual participants, watershed groups try to change the attitudes of community members. CWOs want to "problematize" decisions and behaviors at both the individual and community level: "before we started I'm not sure anybody knew what a watershed was.... we have signs all over the area ... saying you are now entering ... Watershed Area. Please don't do this and don't do that.... they now think more about what's going into the lake...."

These individual and community attitude changes are not as easily measureable as in-stream changes (such as pH or macroinvertebrate populations). However, for many CWO leaders, changing individual behaviors is the key to long-term ecosystem health.

Unintended Outcomes: Building Local Capacity

Positive unintended outcomes (not stated goals of the group) can occur at multiple levels, including individuals participating in the group, the organization itself, and the larger community.[17] In aggregate, these can lead to significant social change surrounding environmental issues. CWO members' actions have multiplier effects on their communities in ways they may or may not recognize.

Building individual capacity. CWO leaders report that they and other members have increased their knowledge of natural resources and environmental disciplines, including ecology, chemistry, and biology. They attend trainings to increase their capabilities for watershed work. As a result, they have become experts on the local environment and, consequently, assets to their communities.

Group leaders develop skills in communication, negotiation, and partnership development. They also gain skills related to political advocacy, resulting in an increased ability to effect change in their communities: "Ann and Sarah[20] have become these powerful two women. They have strong political ties now. They have written letters and letters and have so much money in their watershed from doing all this work themselves.... They have the sense that there is something we can do about our problem...."

---

[20] *Names changed to protect confidentiality.*

Individuals also report increasing their personal networks and forming relationships with people within their community. These relationships offer social support and friendship – essential to creating a feeling of belonging within a community – as well as networks of people they can mobilize when needed for watershed action or other community work.

Building CWO capacity. The continued existence of an organization can be an accomplishment in and of itself.[21] Member turnover within CWOs can be high, as leaders burn out or move away. "I think [a success is] still being a strong group. Because we could have went by the wayside.... DEP ... said ... you guys will be gone in a year."

A few watershed group leaders describe success as developing elements needed to sustain the organization, such as membership, leadership, or specific organizational skills among their members and/or staff. However, relatively little attention was paid to these issues; CWO leaders seemed more interested in accomplishing on-the-ground projects than in investing in these organizational management processes.

Watershed group leaders recognized their impact on the development of other conservation groups: "In the last year, I have been to 3–4 sportsmen's groups meetings ... the whole meeting is about how to conserve the stream.... I think [our activities] really gets more of the local community on board thinking conservation and preservation of what we have here." Watershed group leaders also reported efforts to create multiwatershed or regional organizations addressing issues that cross political or administrative boundaries, such as land use, transportation, or governance.

Building community capacity. Partnering – developing relationships with external organizations – is an important indicator of effectiveness.[22] Consistent with their emphasis on attitude change, CWO leaders believe that collaboration and cooperation will result in more environmental action than other methods (e.g., protesting, litigation): "Our motto is 'conservation through cooperation'."

However, what CWO leaders often do not recognize is the increase in collective capacity these relationships can create. CWOs use the relationships built with other environmental groups, agencies, community groups, businesses, and landowners to facilitate discussion of local environmental problems and seek collaborative solutions. These discussions also identify and prioritize environmental projects, forming stakeholder groups that can successfully attain project funding. These projects are more likely to be effective because multiple stakeholders have participated throughout the process.

The key to partnering is maintaining positive relationships with community groups that have different interests and stances on environmental issues. Successful CWOs describe themselves as "active environmentalists, not environmental activists."

---

[21] Gamson, W. 1975. *The strategy of social protest.* Dorsey, Homewood, IL.

[22] Lee, B. 2005. *Pennsylvania community watershed organizations: form and function.* Doctoral dissertation, The Pennsylvania State University; Stedman, R., K. Brasier, J. Weigle and F. Higdon. 2009. "Cleaning up water? Or building rural community? Community watershed organizations in Pennsylvania." *Rural Sociol* 74(2):178–200.

Three watershed groups specifically spoke of maintaining a reputation of neutrality to facilitate community dialog and projects with diverse participants: "...we tell them in a way that they can hear it.... it ... gives us the opportunity to function with a bunch of different groups ... because we aren't telling you you're wrong. We're not ... making judgments on what somebody is doing."

Building political capacity. Many CWO leaders reported serving as skilled volunteers on government boards, committees, and advisory groups. Other groups send organizational representatives to local government meetings to provide formal comment: "You really can influence the municipal leadership ... if you go to the meetings, properly informed with good facts...." These interactions were generally described as nonconfrontational but persistent (i.e., "bugging" them).

As a result of their level of organization, their activities, and their nonconfrontational tactics, some CWOs report a change in local politics and decision making. "[W]e were a squeaky wheel and we did get the grease because we were organized.... If you get people clamoring the politicians will listen and that's what I think the role of a watershed [organization] is...." Generating publicity is a key tactic for CWOs, including letter-writing and phone campaigns to environmental agencies, local government, and local media; providing educational material during local elections; and holding public events and meetings. "Well, we speak out! We're not shrinking violets!" Similar publicity tactics are used to influence decisions at the state level.

When publicity efforts fail, a few groups reported working to change local government itself: "...we had the impression that they [township supervisors] pretty much ran things the way they wanted and we felt strongly that the process of democracy needed more citizen involvement and a more dynamic exchange between our leadership and the people and we had to go to court to press some of our thoughts...."

Agency Participation in Pennsylvania's Community Watershed Organizations

The agency role within CBRM organizations varies considerably. In formal stakeholder partnerships, agencies are prominent partnership members. In the voluntary, citizen-based organizations primarily described here, agency staff members participate as leaders, members, technical support, and organizational support (including assistance to obtain and manage finances and providing meeting space and resources for mailings).

It is important to consider how agency personnel participation might influence watershed organizations.[23] The technical skills, knowledge, and resources of

---

[23] Bidwell, R.D. and C.M. Ryan. 2006. "Collaborative partnership design: the implications of organizational affiliation for watershed partnerships." *Soc Nat Res* 19:827–843; Moore, E.A. and T.M. Koontz. 2003. "A typology of collaborative watershed groups: citizen-based, agency-based, and mixed partnerships." *Society & Natural Resources* 16(5):451–460.

agency personnel can be crucial for CWO activities. However, agency resources may be limited, and technical staff may lack the time, training, and experience to help manage organizational processes.[24] Agency personnel participation can influence watershed groups' choice of projects, focusing on those important to existing agency stakeholders instead of those that might be identified through a more inclusive process.[25] Projects that incorporate the experiences, opinions, and values of all stakeholders have greater legitimacy in the public eye and a higher likelihood of success.[26] Consequently, it will be important for agency personnel to identify ways to participate in CWOs that support the unique capacity-building role of CWOs and that foster an inclusive decision-making process.[11]

Many Pennsylvania CWO leaders mentioned one person from governmental agencies – the county's watershed specialist. Watershed specialists are county conservation district employees paid through grants from the state's Department of Environmental Protection (DEP). As of 2003, more than three-quarters of Pennsylvania's counties had a watershed specialist. CWO leaders also mentioned participation of other personnel from DEP and conservation districts, as well as other governmental agencies such as the Natural Resources Conservation Service.

In Pennsylvania CWOs, agency personnel play all the roles described above. They helped form the organizations, serve on boards, and, in some cases, hold the organizations' checkbooks. In trying to understand the agency impact on CWO outcomes, we identified those organizations for which agency personnel played an instrumental role in the formation of the CWO. Nine of the 28 CWOs fit this description.

In a comparison of these two groups, the only outcome affected by this sponsorship is local decision making. Of the nine agency-affiliated groups, only two report that their CWOs have directly influenced local natural resource decision making. The agency-affiliated groups tended to see themselves as the neutral facilitator, with less awareness and/or interest in influencing local decision making through political action. "I think the whole philosophy for [our group] is that we are the group in the community that tries to bring all of [our region] to the table to discuss environmental issues.... we feel that ... we need to be the ones that keep everybody talking." These groups were more likely to emphasize their ability to facilitate community dialog over complex environmental issues and partnering with local governmental units.

---

[24] Leach, W.D. and N.W. Pelkey. 2001. "Making watershed partnerships work: a review of the empirical literature." *J Water Resour Plann Manag* 127(6):378–385.

[25] Bidwell, R.D. and C.M. Ryan. 2006. "Collaborative partnership design: the implications of organizational affiliation for watershed partnerships." *Society & Natural Resources* 19:827–843.

[26] National Research Council (2008) "Public participation in environmental assessment and decision-making." In: Dietz, T., P.C. Stern, Committee on the Human Dimensions of Global Change, Division of Behavioral and Social Sciences and Education (ed) *Panel on public participation in environmental assessment and decision-making*. The National Academies Press, Washington, DC.

Implications for Working with Community Watershed Organizations

CBRM efforts arose out of the need for an alternative to the command-and-control policies of federal and state governments. As such, they fall into the category of civil society – that social space between government and the private sector. CBRM efforts are directed by citizens living within the watershed, filling the niche between what government and private businesses and individuals can or are willing to do. In practice, however, CBRM organizations often work in partnership with both governmental agencies and private organizations – and need to do so to be effective. To maintain the citizen-led nature of these organizations, it will be essential for leaders and professionals to seek community-based (rather than agency-based) participation and direction. At the same time, CBRM organizations must access technical and funding information needed to sustain their watershed protection and remediation efforts.[27] This collective effort of both public agencies and volunteers can be effective in attaining both environmental improvements and increased social capacity.

However, this effectiveness is tempered by a number of factors that those who work with CBRM organizations need to address since these issues have impacts at the local, regional, state, and federal levels. The first is the role of governmental agency professionals. Their support and technical expertise are essential to successful watershed management. However, agency professionals are not necessarily trained in developing and managing citizen-led, collaborative, consensus-based efforts. Additional training in these areas can increase the effectiveness of agency participation, as can inclusion of social science professionals in the mix of public support staff for watersheds.

Second, the topics CBRM groups choose to address significantly impact their level of effectiveness. Most groups discussed here focus on direct environmental remediation and attitude change. However, these approaches are limited. They do not address systemic issues, such as land use planning, which contribute to watershed degradation. As groups move toward these issues, they may find their voluntary, consensus approaches challenged by a need for greater activism and/or the use of different tactics (such as political advocacy or litigation).[13]

Third, the capacity of CBRM organizations is limited by their lack of recognition of the importance of internal organizational development and decision-making processes, such as building trust, facilitation, strategic planning, collaborative problem solving, financial management, leadership development, and volunteer management. The number of core members is usually small (less than 25). Leaders serve with much passion and energy, but very few organizations have leadership recruitment and succession plans to avoid "burn-out." CWO leaders in our study

---

[27]Morton, L.W. 2003a. "Civic watershed communities: walking toward justice: democratization in rural life." *Res Rural Sociol Dev* 9:121–134; National Research Council (2008) "Public participation in environmental assessment and decisionmaking." In: Dietz, T., P.C. Stern, Committee on the Human Dimensions of Global Change, Division of Behavioral and Social Sciences and Education (ed) *Panel on public participation in environmental assessment and decision-making*. The National Academies Press, Washington, DC.

generally did not recognize organizational development as an important outcome in and of itself. Professionals working with these groups need to recognize the relationship between developing and maintaining a strong, focused organization and the organization's ability to provide long-term positive environmental and social impacts, and convey this recognition to CWO leaders as well as other agency personnel.[28] In so doing, they can increase awareness of the need for organizational development, and help the CWO seek such training.

Fourth, funding is a perpetual problem for voluntary organizations. Although competitive funding opportunities exist for specific projects, these often do not support related activities, such as organizational management, development of projects, and development of partnerships. These other activities are equally important to the environmental and social outcomes produced by these groups. Funding for basic organizational management might be difficult to obtain from governmental organizations but could be a role for the philanthropic community. Further, by linking organizational strength to both the intended and unintended consequences, new funding opportunities might arise through collaborations of community development and environmental agencies/organizations.

Finally, the unintended impacts identified in this analysis – on individuals, organizations, and the community – often go unrecognized by CWOs and their partners. Consequently, these outcomes are not celebrated, nor are they seen as potential goals CWO leaders could strive to achieve. Further, the importance of these tools for addressing systemic problems (such as land use) is not understood, nor is their potential for creating new, integrative funding and technical resources. Professionals working with such organizations should recognize the long-term impact of such unintended social consequences and incorporate the development of these impacts in organizational planning as both intended outcomes and means to achieve other environmental outcomes.

CBRM can have profoundly positive impacts on the ecosystem and the surrounding community. Our findings suggest that both intended and unintended consequences of CWOs can lead to changes on the landscape. Both types of outcomes should be recognized. In particular, CWOs and their partners should seek to make the unintended intended – incorporate the capacity-building outcomes into their planning, assessment, and reporting. Additionally, they can link organizational development with environmental outcomes, which can potentially create new funding and collaborative opportunities. Strengthening the organization increases its ability to achieve environmental protection as well as increases its capacity for future action. With the proper support, CWOs can be a key mechanism for enhanced public interaction, participation, and governance within watersheds and their communities.

Acknowledgements We gratefully acknowledge funding from the Center for Rural Pennsylvania for this project.

---

[28]Born, S.M. and K.D. Genskow. 2000. *The watershed approach: an empirical assessment of innovation in environmental management.* National Academy of Public Administration Research Paper Number 7. http://www.napawash.org; Leach, W.D. and N.W. Pelkey. 2001. "Making watershed partnerships work: a review of the empirical literature." *J Water Resour Plann Manag* 127(6):378–385; Morton, L.W. 2003a. "Civic watershed communities: walking toward justice: democratization in rural life." *Res Rural Sociol Dev* 9:121–134.

Chapter 12
Community Watershed Planning: Vandalia, Missouri

Daniel Downing, Robert Broz, and Lois Wright Morton

Impaired Waters

The United States Environmental Protection Agency's (EPA) nine-element framework for watershed planning sets the regulatory expectations when a watershed is classified as impaired and is required to develop a Total Maximum Daily Load (TMDL) plan.[1] However, before regulatory requirements can be met, watershed planning must begin with citizen awareness, acceptance of the impaired classification, and the building of partnerships. Involving citizens in TMDL development is time consuming but necessary if land use practices are to change so that impairments are reduced and water bodies are restored to meet their designated uses.

In 1994 the reservoir that supplied Vandalia's public drinking water registered atrazine levels in excess of 80 parts per billion (ppb), far above the maximum of 3 ppb for treated drinking water regulatory standards. The local community recognized this as a significant issue and mobilized to solve the problem. Five years after the establishment of a local watershed management committee and the writing of a watershed management plan (a plan that farmers and community members supported), Vandalia's untreated water averages less than 6 ppb of atrazine in the raw water. At this level, the Vandalia water plant can effectively and affordably treat its water to meet EPA's 3 ppb drinking water standard.

This chapter uses the Vandalia experience to take the reader through developing a local citizen group and applying the EPA planning and implementation process to accomplish the nine minimum elements of a watershed plan. Although the plan developed by the Vandalia community predates the current nine-element guidance from EPA, we present the planning process to illustrate where these elements fit.

---

[1] Handbook for Developing Watershed Plans to Restore and Protection our Waters. 2005. USEPA (http://www.epa.gov/owow/nps/cwact.html).

D. Downing (✉)
Agricultural Extension-Food Science & Nutrition, University of Missouri, 226 Agricultural Engineering Building, Columbia, MO 65211, USA
e-mail: downingD@missouri.edu

L.W. Morton and S.S. Brown (eds.), *Pathways for Getting to Better Water Quality: The Citizen Effect*, DOI 10.1007/978-1-4419-7282-8_12, © Springer Science+Business Media, LLC 2011

Citizen roles are discussed and illustrated with quotes from nine interviews with public officials, water plant managers, Natural Resource Conservation Service (NRCS) and extension agency staff, community leaders, and landowners. Data are drawn from multiple sources, including public records, newspaper articles, and audio recorded interviews with local residents in September 2004. Interviews were transcribed word for word and researchers reconciled concepts and identified the following key themes: agriculture vs. urban; atrazine impairment; conservation practices; conservation ethic; technical assistance; relationships; water management; and nonatrazine water quality issues. NVIVO software was used to analyze main themes and develop a table of quotes in support of major themes.

Watershed Planning and EPA Nine Elements of a Watershed Management Plan

The EPA defines an impaired waterbody as "a waterbody that does not meet criteria that support its designated use. The criteria might be numeric and specify concentration, duration, and recurrence intervals for various parameters, or they might be narrative and describe required conditions such as the absence of scum, sludge, odors, or toxic substances."[2] Classification as impaired places it on the 303(d) list defined in Section 303 of the U.S. Clean Water Act. Once on the list, the state must produce a restoration target document called a TMDL. A TMDL focuses on specific segments of a waterbody, sources, or pollutants. A watershed protection plan goes beyond the TMDL-specific pollutants and segments of a waterbody to include the entire watershed.

Table 12.1 lays out the watershed planning process and identifies the nine elements of a TMDL that should be incorporated into a planning process. The process begins with building partnerships and relationships in the watershed prior to preparing and implementing the TMDL document. The core of the planning process includes characterizing the watershed, setting goals and identifying solutions, and designing and implementing a restoration/management plan. The last steps make explicit the expectation that adjustments will be made to the plan as progress is measured and evaluated. A good watershed protection plan not only responds to the specific impairment but also identifies potential future threats, pollutants, habitat, and restoration issues and develops a holistic plan beyond the immediate threat. Further, the plan is iterative and adaptive using a cycle of planning, implementation, and evaluation, with revisions to the plan as conditions change. It is important to remember that watershed planning is not an exact science,[3] it occurs in a dynamic environment. There is considerable variability among watersheds related to the impairment, land uses, and the socioeconomic relationships among those who live and work in the watershed.

---

[2] P. 2-2 Handbook for Developing Watershed Plans to Restore and Protection our Waters. 2005. USEPA (http://www.epa.gov/owow/nps/cwact.html).

[3] P. 1-7 Handbook for Developing Watershed Plans to Restore and Protection our Waters. 2005. USEPA (http://www.epa.gov/owow/nps/cwact.html).

Table 12.1 Incorporating EPA (Environmental Protection Agency) TMDL (Total Maximum Daily Load) nine elements into watershed planning

Build partnerships	
Characterize the watershed	
Element #1	Identify causes and source of pollution that need to be controlled
Finalize goals and identify solutions	
Element #2	Determine load reductions needed
Element #3	Develop management measures to achieve goals
Design an implementation program	
Element #4	Develop implementation schedule
Element #5	Develop interim milestones to track implementation of management measures
Element #6	Develop criteria to measure progress toward meeting watershed goals
Element #7	Develop monitoring component
Element #8	Development information/education component
Element #9	Identify technical and financial assistance needed to implement plan
Implement watershed plan	
Measure progress and make adjustments	

Handbook for Developing Watershed Plans to Restore and Protect Our Waters, 2005, pp. 2–17. USEPA. http://www.epa.gov/owow/nps/cwact.html

Vandalia, Missouri

Vandalia, Missouri, population 2,500, like many small towns in northern Missouri, is reliant upon surface water for its public water supply. The area economy relies heavily on agriculture and supporting agribusinesses. The primary water source for Vandalia is a 28-acre reservoir supplied by a 3,638 acre watershed. The town of Vandalia is located in Audrain County, whereas the municipal reservoir and its watershed are located in neighboring Pike County. Since the town of Vandalia was in a different county than the watershed, many of the landowners had little association with the town and may not have seen the importance of their role in helping protect the town's water supply.

Much of the acreage draining into the Vandalia reservoir is used for production of corn and grain sorghum. Although used in production of other crops, in the Midwest, atrazine is the herbicide most commonly applied in the production of corn and grain sorghum. Atrazine is highly soluble and therefore is easily transported from agricultural fields into public waterways and eventually to surface water catchments, like Vandalia reservoir, where it can pose an ecological and human health threat. Although producers in the watershed had previously implemented some form of recommended conservation practices, their focus had been on erosion control rather than reduction of atrazine runoff.

Herbicide manufacturers worked closely with EPA, monitoring raw water and finished water, and these data were used by EPA in setting the 3 ppb maximum standard. Monitoring results showing elevated pesticide levels in the runoff were shared with local farmers, causing them to shift their conservation efforts toward runoff control.

Impaired Water Classification – 303(d) Listing

In 1997 Vandalia's water plant manager was alarmed to see elevated levels of atrazine in the untreated water supply. The chief water operator at the time says, "...they're taking it right from the surface near the intake at the reservoir, so this is raw water without any treatment ... this is triazine ... there's about four triazines, one of them being atrazine... ." He continues his story, "In '96 we started with the volunteer Novartis sampling plan, where we sent in samples of our raw water and finish water... and they would analyze it for us and give us the results of the atrazine level. Well, in '97 in May, the lab called at the water treatment plant and informed us that we were well over the limits. We were 85 parts per billion of atrazine... ."[4] The treatment process used at Vandalia could remove low levels of atrazine but to reach the maximum allowable contamination level of 3 ppb in the finished water would require additional treatment and substantial expense. The plant manager contacted the Missouri Department of Natural Resources and was able to borrow some equipment to assist with the treatment and removal of atrazine. It was realized that the extremely high atrazine levels in the raw water could not be removed from the water supply even with additional treatment. This situation triggered major concern by city officials.

Building a Partnership and Characterizing the Watershed

Recognizing that they had little control over the water coming into the reservoir, the water plant manager met with the city manager to discuss possible options for reducing the atrazine load. Getting the farmers and landowners in the neighboring county to acknowledge the concerns and to realize that they may be partially responsible for the problem became the next challenge. Local landowners needed an opportunity to understand the concern, take some responsibility, and contribute to the solutions. One farm cooperator whose crop lands drain into the Vandalia drinking water reservoir says, "[it] has always been by far the most economical chemical. Of course, it was part of the watershed,

---

[4]Line127–131 1997 chief operator of Vandalia water plant, Vandalia, MO interviews September 2004.

atrazine problem ... the city was interested in the runoff rate reduced so they could back off on carbon ... ."[5]

Another farm cooperator talks about atrazine recommendations, "we put a test plot in and demonstrated ... Dad was still active in farming [so it was some years ago] ... at that time, they were promoting large amounts of atrazine, maybe three pounds [per acre], and so it needed to be reduced."[6] Weather is an important contributor to excess atrazine runoff, "...if a chemical was put on there real early, and the rains came..."[7] it was lost in the sediment erosion process.

Community and Technical Assistance

University of Missouri Extension was asked to assist the water plant manager in informing the local producers of the problems that had been identified. At about the same time, several of the local producers contacted the University of Missouri Extension Water Quality office with concerns about their role in the situation. Through a series of one-on-one meetings and informational letters, farmers were made aware of the concerns. A plan was devised to work directly with local producers and municipal officials to see where there was a common interest.

Because the producers were already involved in many conservation efforts, they wanted to know what else could be done to prevent the herbicide runoff. This was a group of concerned, open-minded producers that had the desire to do something, but did not know how to make the change come about. Vandalia's mayor says, "We had a document showing this many parts per billion of atrazine, so yeah, that wasn't questioned. The question was how to get it back down."[8]

The water plant manager, the city administrator, and University of Missouri Extension met to discuss possible ways to engage the producers in working with city officials. They discussed what they wanted to accomplish, not placing the blame on anyone in particular, what was the cost of treating the water before it could be released, and what were long-term strategies that might help reduce atrazine loading. They developed a strategy to meet with key players in the region and formed a Watershed Management Committee. This included personal visits, phone calls, and personal invitations to a series of meetings where the atrazine issues were discussed. Although there are only 14 families that farm in the watershed, there were in excess of 50 individuals with property interest in the area. The decision was made to focus on the farm operators and to guarantee their attendance at the meetings by making personal contacts with each producer.

---

[5]Line 129–138 Farmer #1 Vandalia, MO interviews September 2004.

[6]Line 212–215 Farmer #2 Vandalia, MO interviews September 2004.

[7]Line 216–224 Farmer #2 Vandalia, MO interviews September 2004.

[8]Line 370–373 Mayor Vandalia, MO interviews September 2004.

Agriculture vs. Urban Interests

Extension representatives started working directly with the water plant manager to set up meetings to get the farmers and local town people together and discuss the concerns. At the first general information meetings, there was a divide between the farmers who lived and worked in the watershed and those from the town of Vandalia who drank the water. Farmers were unconvinced that they should carry the whole burden of cost, "we spent more on changing our chemical program than they ever spent on putting carbon in the city water, but it depends on whose pocketbook it comes out of, whether it is perceived as costly or not."[9] The perception from the town side was that the farmers were over applying and that was creating the problem. The farmers were applying at label recommendations (legal rates) and had invested capital in management practices and found it difficult to understand how they were causing the problem.

Setting Goals and Solutions

Three public meetings were held in the beginning process to identify what local stakeholders thought were the issues and to discuss management practices that producers were willing to consider. At each of these meetings, the local residents from both the rural watershed and the town were asked to help identify what their goal was for the watershed, to help outline possible ideas on what could be done, and what role they were willing to play. The mayor and the water plant manager explained what the city was doing at the plant to correct the problem. They explained the need for help from the local producers in reducing the levels in the raw water.

At the meetings, producers and town residents tried to keep a positive tone and to discuss what each group could do to protect water quality in the reservoir. It was recognized that the producers needed to make a living off the land and that agriculture was a major economic base for the community. The partnership between the agriculture community and the city had to allow the farmers to continue farming and still provide the city with a suitable water supply. Several different water quality issues were identified at these meetings, including sediment, atrazine, and nutrient loading.

Relationships

Once started, cooperation between the producers and the area residents was high. The chief water operator recalls talking with the city administrator and the mayor about

---

[9]Line 132–138 Farmer #1 Vandalia, MO interviews September 2004.

their responsibility, "...I really want to stress this ... first off, we aren't going to go in and point fingers. We aren't going to go in ranting and raving. So what I presented them at the November meeting was a breakdown ... the cost of producing a thousand gallons of water, what it used to be, and then with the added carbon cost, the treatment cost ... we cut the bill in half. Instead of pumping 500 gallons a minute, we slowed down that pump to where we could actually treat the water with a higher dosage... ."[10] The chief operator was a key person in setting the tone of public meetings, "Instead of just going to them and saying, 'well, how are you going to fix this,' we let them know what we were doing to try to correct this problem ... we went to the producers and said, 'you know, this is what we're doing, and we'll need some help'."[11]

The situation in Vandalia was similar to those in other communities that rely on surface water at risk of contamination from agricultural runoff. The difference in Vandalia was that a group of farmers acknowledged their share of the responsibility and were willing to do something about it. One farmer says, "...people ... can work with other people, even if they don't have exactly the same opinion."[12] He adds, "I can't say enough about the town of Vandalia how they were very cooperative in working to help get ... grants and so forth."[13]

At each meeting, sign-in sheets were sent around asking people to identify if they would be interested in serving on a steering committee to develop a watershed management plan. Once a steering committee of interested producers and residents was established, meetings were held monthly to bring all parties to a common level of understanding. Management practices and the overall vision or outcome were determined by the local residents. The steering committee started working on the issues that were identified at the three public meetings but focused on atrazine levels as their primary concern. Many conservation practices were already in place, but other cost-effective practices were identified that could still be implemented utilizing existing cost-share programs. University Extension facilitated the steering committee meetings when needed and helped lead the group in identifying other agency partners that could provide financial and technical support. Guest speakers were brought in to discuss possible activities that would help increase awareness in the community and to keep people actively involved.

Designing and Implementing a Plan

Agency partners, including the NRCS, Missouri Department of Agriculture, Missouri Department of Natural Resources, and Missouri Department of Conservation, attended many of the meetings and were able to offer financial and technical assistance to farmers willing to try new and different practices that would

---

[10] Line 204–213 chief water operator Vandalia, MO interviews September 2004.

[11] Line 235–239 chief water operator Vandalia, MO interviews September 2004.

[12] Line180–186 Farmer #2 Vandalia, MO interviews September 2004.

[13] Line 133–134 Farmer #2 Vandalia, MO interviews September 2004.

protect water quality. The NRCS specialist estimated that approximately 1,700 acres of cropland and 1,300 acres of grassland needed to be evaluated and assessed.[14] One hundred percent of the producers in the watershed signed agreements to participate in water quality improvement practices. Each producer recognized that to be part of the community meant doing what they could to protect the community's water supply. Atrazine applications were reduced to less than a pound of active ingredient with more attention paid to weather conditions and spring run-off and sediment loss.

In many cases, the best identified practices were when farmers talked across the fence to see what others were doing that seemed effective. A farmer reports, "...one of us would have corn up here, we'd all have corn up here. So we tried to get on a rotation where we wouldn't all have the same thing, I think that's one thing that helps. We used some alternative chemicals."[15] Different practices and combinations of practices were used to mitigate atrazine runoff including:

- Timing of application as a management option for farmers. Even with a narrow window of time to get planting done, the farmers now realize that as part of the watershed community they need to be aware of how their decisions can affect others who rely on the Vandalia reservoir for their daily drinking water.
- Staggered crop rotations. By having no more than 60% of the tillable acres in corn or grain sorghum each year, the available amount of atrazine applied in any 1 year is reduced.
- Grass filter strips. Filter strips can tie up and degrade some of the atrazine that is found in the run-off water.
- Integrated pest management practices. These practices target specific weed and insect populations and site-specific application that reduces the levels of agricultural chemicals detected in the runoff.
- Environmental Quality Incentive Program (EQIP) of NRCS – includes a structured series of environmentally friendly practices eligible for incentive payments through USDA if implemented by private landowners.
- Missouri Conservation Reserve Enhancement Program (MoCREP) takes marginal lands in drinking water watersheds out of production and establishes them in grass for run-off and erosion control. The MoCREP program expands the CRP program and is targeted at sensitive watersheds that supply water to public reservoirs.

Collectively these management strategies have kept the atrazine component of the runoff to treatable/manageable levels.

---

[14]NRCS Conservationist for Pike and Lincoln counties September 2004.

[15]Line 84–87 Farmer #2 Vandalia, MO interviews September 2004.

Measuring Progress and Making Adjustments

Since the establishment of the watershed management committee and the writing of a watershed management plan, the levels of atrazine in the reservoir decreased from a peak of 85 ppb in the untreated water to averaging less than 6 ppb in the raw water. At this level, the water plant manager can effectively treat the water without additional expense. There are still some short spikes or elevated levels due to weather patterns and planting times, but the overall watershed management practices are now a component that many of the farmers build into their farm planning process.

From its beginning, it took the Vandalia Watershed Committee 3 years to form and reach synergy before they could function as a group and fully implement their watershed management plan. Numerous meetings, field days, educational events, and media coverage were needed to bring the people together and keep them informed of the progression of the watershed planning process.

Throughout that time, monitoring of the raw and finished water revealed the effectiveness of changing agricultural practices as levels of atrazine in the reservoir declined. Agency personnel worked diligently with producers to find management practices that could be cost shared and to provide technical support in placement and operation of practices. For their efforts, the Vandalia Watershed Committee won the Missouri Chamber of Commerce's Governor's Award for Environmental Stewardship.

The project was initially driven by the concern of health implications and potential regulatory action. However, ultimately Vandalia's success was built and sustained when town and rural residents joined together to address the threats that impacted both segments of the community. The message that the local farmers need to protect the city's water supply is being passed on to new producers who lease and buy land by those already farming in the area. The future challenge will be whether producers will continue to be vigilant and sustain practices that protect Vandalia's drinking water source.

The Watershed Plan Process

The development of an implementable watershed plan and response to federally required TMDLs for impaired waters cannot be accomplished by technical professionals alone. Nor do local citizens have the knowledge, resources, or capacity to respond alone. A workable plan requires a partnership among watershed citizens and technical experts. A local farmer says the community could not have responded without the external catalyst of Missouri Extension, "...don't think it would have had the structure it did. Because there were times when [extension specialist] might not come or he'd be late, and we were a little bit lost, so I think [the community] needs [external support] ... they know what's going on other places, things that we possibly could do."[16] It takes a merging of resources.

---

[16] Line 342–359 Farmer #2 Vandalia, MO interviews September 2004.

As of this writing, the Vandalia community continues to maintain compliance in their finished water with the safe drinking water standards. Their highest raw water reading in the past 5 years has been a one-time detection of 12 ppb. They are still enrolled in a voluntary monitoring program. Vandalia's watershed has been used as one of the reference watersheds through EPA for decisions on atrazine management with recommendations to increase the frequency of sample monitoring. The watershed committee continues to meet on an as-needed basis and is currently revising their watershed management plan to more directly address EPA's nine-element criteria.

Dealing with nonpoint source pollution requires a dramatically different approach than regulation of point source pollution. Nonpoint source pollution is much more complicated, involving many more personalities and major issues relating to private property rights and voluntary compliance. The success of farm and urban communities in addressing atrazine levels in the Vandalia reservoir shows how these issues can be negotiated by the development of common goals and effective partnerships among citizens and agencies.

Chapter 13
The Role of Force and Economic Sanctions in Protecting Watersheds

Kristen Corey and Lois Wright Morton

Decisions about natural resource functions, their value to society, and their management are made by governments at multiple levels. Public environmental policies are a mosaic of local city and county, state, and federal laws, agency regulations and mandates, historical and customary practice, economic incentives and sanctions, as well as legal interpretation. Flora's Agroecology Management Model (see Chap. 2) labels these strategies "force," which apply positive and negative pressures in a top-down manner to obtain compliance in managing resources for the public good. Resource management using a top-down approach, through legislative force, economic incentives, and sanctions, can be efficient and is often the arbitrator of competing self-interests of societal sectors. Citizen involvement under these conditions is formalized through public hearings, formation of special interest lobby groups, and representation by elected officials.

Central to decisions about natural resource management are public values, beliefs, and knowledge about the environment in general and about specific natural systems such as watersheds. These public views must somehow be linked with the dynamic and changing sciences of ecosystem relationships and the application of technologies and practices that utilize, protect, and/or restore the natural resource base. Not unexpectedly, the complexity of managing natural resources from an ecosystem perspective leads to conflicting viewpoints and challenges to public and private decisions. Planning departments and voluntary citizen boards who develop comprehensive land use plans and land use development guidelines are part of city, town, and county governments' strategies to ensure that citizens' interests and preferences are represented.

People within a single watershed may have different experiences, knowledge, and beliefs about the function of water, the appropriate uses of water, the extent of water pollution problems, and whether their water resources are at risk. The conservationist, the environmentalist, the industrialist, the agency official, and the farmer often do not agree on how this resource should be allocated, or on how and whose responsibility it is to protect it.

K. Corey (✉)
Iowa State University, 313 NW Bramble Road, Ankeny, IA 50023, USA
e-mail: kcorey@dhs.state.ia.us

L.W. Morton and S.S. Brown (eds.), *Pathways for Getting to Better Water Quality: The Citizen Effect*, DOI 10.1007/978-1-4419-7282-8_13, © Springer Science+Business Media, LLC 2011

We present the Eastern Nebraska Saline Wetlands surrounding Lincoln, Nebraska as a case study to illustrate the experience of planning and zoning staff and public environmental agencies in protecting water resources. City and county planning boards, other agencies with environmental protection missions, and The Nature Conservancy (TNC) have forged a partnership to protect the disappearing saline wetlands ecosystem using the tools provided by statute and public policy. Suburban sprawl into farmland, concern about property rights, and a fragmented sense of community are social factors that have made it difficult to build a common vision and implement science-based land use practices available to protect and restore the wetlands. Citizen input to the process has been primarily through public hearings and established not-for-profit organizations representing environmental protection and restoration goals. Many individual citizens have challenged public agencies' actions and the invocation of the Endangered Species list has been marked with some controversy.

Interviews with farmers and other landowners, planners, the Nebraska Department of Environmental Quality (NDEQ), and other conservation agencies' staff offer insights into the protection of the saline wetlands ecosystem as viewed by different conservation partners and the intended and unintended consequences to landowners. These viewpoints also reflect how the application of the Federal Endangered Species list has influenced the process of integrating diverse local environmental priorities.

Local Conflict over Property Rights

Diverse viewpoints on property rights related to environmental protection add to the complexity of land use planning and environmental goals and actions. As in most private land and natural resource debates, arguments over the rights that private property owners have to their own lands can lead to intense and sometimes contentious public discourse. Anderson and McChesney[1] write that:

> Property, in its most complete form ... gives its owner the right to derive value from the asset, to exclude others from using it, and to transfer the asset to others.... However, property rights may be less complete, allowing an owner to derive only some value from an asset, exclude only some people from using it, or transfer only certain uses for a specified time period.

In some cases, property owners may not have complete access to or control over all of their assets, especially when it comes to natural resource and water issues. Since many natural resources are considered "public goods," the presence of this type of good on one's so-called private property can present conflicting interests. The line between a common and an individual good can be wide open to interpretation. Landowners lean toward their ability to do what they want with their lands, including

---

[1] Anderson, T. L. and F. S. McChesney. 2003. "The Economic Approach to Property Rights." Pp. 1–11 in *Property Rights: Cooperation, Conflict and Law*. Princeton, NJ: Princeton University Press.

the opportunity of developing if they wish[2] and agency professionals (based on legal mandates) tilting toward preservation or conservation for the common good. Jackson-Smith et al.[2] write that, "restrictions on land development are frequently met with intense political opposition from landowners who resent having their 'development' options limited and worry about the reductions in the market value of their property."

Since limitations on private property rights are generally not met with enthusiasm from most landowners who wish to sell their land for development, one can imagine that these situations are usually not taken lightly. A struggle between local landowners and local and Federal agency representatives often surfaces as private landowners who wish to develop their land are forced to choose the common good over their own individual preference.

The Disappearance of the Eastern Nebraska Saline Wetlands

Near Lincoln, Nebraska, there is a small area of what are known as the Eastern Nebraska Saline Wetlands (Fig. 13.1). These unique wetlands are home to a number of species indigenous to the region, including the Salt Creek tiger beetle (*Cicindela nevadica* var. *lincolniana*) and the saltwort plant (*Salicornia rubra*), both listed as "endangered" species.[3] In addition to providing habitat for these endangered, native species, the saline wetlands also serve important hydrological, ecological, social, and historical functions to Lincoln and its surrounding areas.[4]

Although the Eastern Nebraska Saline Wetlands serve important social, ecological, and hydrological roles for the city of Lincoln, much of the original saline marshlands present when Lincoln was settled no longer exist. Prior to settlement of the area, the saline wetlands occupied nearly 20,000 acres in the Salt Creek and its drainage basins.[5] Today many of the wetlands have been drained and filled

[2] Jackson-Smith, D., U. Kreuter, and R. Krannich. 2005. "Understanding the Multidimensionality of Property Rights Orientations: Evidence from Utah and Texas Ranchers." *Society and Natural Resources* 18:587–610.

[3] Farrar, J. 2005. "Preserving the Last of the Least: A Partnership of Organizations is Working to Protect the Saline Wetlands." *NEBRASKAland Magazine*. January–February, p. 46; Willey, S. and B. Perkins. 2005. "Service Lists Salt Creek Tiger Beetle as Endangered." News Release: U.S. Fish and Wildlife Service, March 6, 2007. (http://mountain-prairie.fws.gov/pressrel/05-72.htm); U.S. Fish and Wildlife Service, 2005. U.S. Fish and Wildlife Service. 2005. "Salt Creek Tiger Beetle," March 22, 2007. (http://mountainprairie.fws.gov/species/invertebrates/saltcreektiger/index.htm).

[4] Farrar, J. 2005. "Preserving the Last of the Least: A Partnership of Organizations is Working to Protect the Saline Wetlands." *NEBRASKAland Magazine*. January–February, p. 46; Willey, S. and B. Perkins. 2005. "Service Lists Salt Creek Tiger Beetle as Endangered." News Release: U.S. Fish and Wildlife Service, March 6, 2007. (http://mountain-prairie.fws.gov/pressrel/05-72.htm).

[5] Farrar, J. 2005. "Preserving the Last of the Least: A Partnership of Organizations is Working to Protect the Saline Wetlands." *NEBRASKAland Magazine*. January–February, p. 46.

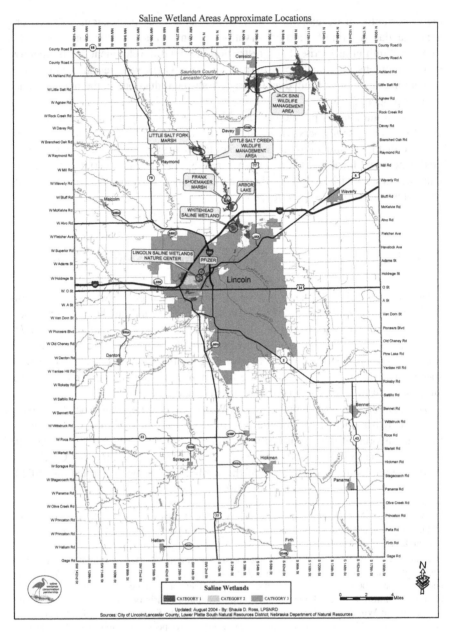

Fig. 13.1 Eastern Nebraska Saline Wetlands (Courtesy of Tom Malmstrom, City of Lincoln/Eastern Nebraska Saline Wetlands Conservation Partnership)

and their hydrology disrupted due to the channelization of streams for both urban and rural uses.[5] Additionally, according to the U.S. Fish and Wildlife Service, "since the late 1800s, more than 90 percent of these wetlands have been destroyed or severely degraded through commercial, residential and agricultural development and transportation projects,"[6] leaving approximately 122 acres of barren salt flat and saline stream edge habitat for the Salt Creek tiger beetle.[7] Of these remaining 122 acres, according to the U.S. Fish and Wildlife service, "merely 15 can be considered 'not highly degraded'."[6]

To put things into perspective, an author from Nebraska Commonwealth who came upon Lincoln in the late nineteenth century described the area in this way:

> Approaching Lincoln from the east, the first remarkable object that meets the eye of the stranger is a succession of what appears to be several beautiful lakes extending along the lines of Salt Creek to the northward and westward of the town, the nearest a mile distant. As their crystal surfaces glisten like molten silver in the sunlight the illusion is complete, and the most critical landscape painter would be deceived as to their character. But there is no water enclosed in their grassy banks…. These apparent lakes are the Salt Basins of Lancaster County, in themselves natural curiosities well worthy of a long journey to visit them. The floor of these basins is hard clay, smooth and level as a brick-yard and polished as that of a Hollander's kitchen. They are covered with a white layer of crystallized salt, wonderfully pure…. Intersecting these salt floors are little streams of salty water, so strongly impregnated that it will abrade the tongue and lips when tasted.
>
> Nebraska Commonwealth, September 7, 1867[8]

Unlike the author of 1867, a present-day visitor to the area would probably describe it using much different terms as both farming and urban expansion have disrupted the hydrology and ecology of this shrinking area.[9] Consequently, a number of local planning and environmental officials have acted to protect these wetlands from further destruction.

Saline Wetlands Conservation Partnership

In 2002–2003, a major step was taken to protect these wetlands through the founding of a local preservation group called the Saline Wetlands Conservation Partnership (SWCP), which emerged out of a $750,000 Environmental Trust grant to the city of Lincoln to preserve and reestablish the saline wetlands.[5] Each of the major partners, including the city of Lincoln, Lancaster County, the Lower Platte South Natural Resources District, The Nature Conservancy, and the Nebraska Game and Parks

---

[6]Willey, S. and B. Perkins. 2005. "Service Lists Salt Creek Tiger Beetle as Endangered." News Release: U.S. Fish and Wildlife Service, March 6, 2007, p. 4. (http://mountain-prairie.fws.gov/pressrel/05-72.htm).

[7]Willey, S. and B. Perkins. 2005. "Service Lists Salt Creek Tiger Beetle as Endangered." News Release: U.S. Fish and Wildlife Service, March 6, 2007. (http://mountain-prairie.fws.gov/pressrel/05-72.htm).

[8]Cunningham, D. 1985. "Villains, Miscreants, and the Salt of the Earth." *NEBRASKAland Magazine*. July, pp. 14–19, 45–47.

[9]Ducey, J. 1985. "Nebraska's Salt Basin: Going, Going, Nearly Gone." *NEBRASKAland Magazine*. July, pp. 20–25.

Commission, initially contributed $75,000 to the cause.[5] Other public sector partners include the NDEQ, the U.S. Fish and Wildlife Service, and the USDA Natural Resources Conservation Service. Nonprofit environmental and other citizen/business organizations in the partnership are Ducks Unlimited, the Nebraska Wildlife Federation, the Nebraska Sierra Club, Lincoln Homebuilders Association, Wachiska Audubon Society, Pheasants Forever, the Conservation Alliance of the Great Plains, and the Cooper Foundation.[5]

The first action the partnership took after the organization was formally established was to purchase land containing saline wetlands from willing sellers. The SWCP also began an effort to get other local landowners to protect their lands from development through conservation easements, which allow the lands to remain in private ownership while still protecting them indefinitely.[4]

Second, the city of Lincoln and Lancaster County approved a comprehensive city plan to guide future growth and expansion that included the goal of preserving the saline wetlands.[5] In the plan, the city and county excluded the majority of the area north of Lincoln from its long-term development plans for the city.[10] In the meantime, with beetle numbers dwindling and urban development rapidly encroaching on critical wildlife habitat, steps were taken by several conservation groups and officials to get the Salt Creek tiger beetle listed on the Federal Endangered Species list.[11]

The Eastern Nebraska Saline Wetlands

Methods and Data

The research data used to develop this case study include in-person interviews and short surveys along with public and private archival data including media reports, the city of Lincoln's comprehensive master plan, and other written public documents regarding the Eastern Nebraska Saline Wetlands and the Salt Creek tiger beetle. We apply an ethnographic methodology to examine the relationships among agency leaders and community members and their attitudes and knowledge about their watershed and the saline wetlands. According to Ellen,[12] "case studies are the

---

[10]LaGrange, T., T. Genrich, G. Johnson, and D. Schulz. 2002. "Implementation Plan for the Conservation of Nebraska's Eastern Saline Wetlands (draft)." Eastern Saline Wetlands Project.

[11]Cochnar, J. and B. Perkins. 2005. "Service Proposes Protection of the Salt Creek Tiger Beetle." News Release: U.S. Fish and Wildlife Service, March 22, 2007. (http://mountain-prairie.fws.gov/pressrel/05-06.htm)

[12]Ellen, R. F. 1984. *Ethnographic Research: A Guide to General Conduct*. London: Academic Press, p. 237.

detailed presentation of ethnographic data relating to some sequence of events from which the analyst seeks to make some theoretical inference."

Fifteen people were interviewed from August 2006–2007. The interviewees were selected based upon four categories representing distinct knowledge, experiences, and perspectives about the Eastern Nebraska Saline Wetlands. Four were chosen for their roles and responsibilities for planning the future of the city of Lincoln metropolitan area. They were three agency staff and one representative citizen: the city's long-range planner, the county planner, the manager of the watershed management section of the public works department, and an appointed citizen planning commissioner.

Six interviewees were selected based on specific environmental planning and land use management responsibilities related to the Lincoln salt marshes and surrounding area. These participants were two representatives from the NDEQ, the District Conservationist for the Lancaster County Natural Resources Conservation Service (NRCS), a representative from the Nebraska Game and Parks Commission, a representative from the Lower Platte South Natural Resource District (NRD), and the director of the Saline Wetland Conservation Partnership.

The third category consisted of two representatives from a not-for-profit environmental organization, The Nature Conservancy, which is a key partner in the Saline Wetland Conservation Partnership. Lastly, seven key landowners with land in the saline wetland designated area or adjacent to the salt marshes were invited for interviews. The names of these landowners were recommended to us by members of the Saline Wetland Conservation Partnership who worked with the local landowners in the area on a regular basis. However, only three of these landowners were interviewed. Two of them graze livestock and own land containing Salt Creek tiger beetle habitat; the third is not a farmer, but owns land adjacent to saline wetlands. Of the four landowners who were not interviewed, two refused and two are farmers who did not return calls and could not be reached for an interview due to planting season responsibilities.

Interviews and Surveys

Interviewees were asked to explain their relationship to the Eastern Nebraska Saline Wetlands; to give a brief overview of the wetlands, and share knowledge and perceptions of the Salt Creek tiger beetle and how it ended up on the Endangered Species list; and then to talk about the community's response to the implementation of the Endangered Species list. All of the interviews except one were audio recorded and transcribed verbatim, and then were coded and analyzed according to common emergent themes.

In addition to face-to-face interviews, each participant was given a short survey that asked whether they strongly agreed, agreed, had no opinion, disagreed or strongly disagreed with the following statements:

- The creeks, rivers and lakes and their surrounding habitats in our community are important because they offer aesthetic, natural places for people to replenish the spiritual side of the human personality.
- The creeks, rivers and lakes and their surrounding habitats in our community exist primarily as resources for people to use to make a living, survive and prosper.
- The degradation of creeks, rivers, lakes and their surrounding habitats in our community must be stopped regardless of the costs to our consumptive economy. These resources are irreplaceable and more valuable than the economic values placed upon them.
- The separation of human activity from the creeks, rivers, lakes and their surrounding habitats in our community is impossible and we must find ways to balance people and nature with the goal of protecting our resources while providing for the economic and social base of our community.

These statements, based theoretically on the Weber[13] categories of preservation, conservation, contemporary environmentalist, and grass-roots ecosystem management, provided insight into respondents' general environmental beliefs. These responses were analyzed in conjunction with personal interview transcripts discussing the Eastern Nebraska Saline Wetlands.

Emergent Interview Themes

Three major themes emerged as interviewees elaborated on the interview questions. First and foremost, a clear division and hierarchy of power and influence was evident among the four different groups involved in this watershed: (1) the planning officials; (2) agency and conservation professionals; (3) The Nature Conservancy representatives; and (4) landowners. Landowners reported a sense of powerlessness and low engagement. Although they were identified as individuals who communicated with the SWCP on a regular basis, they nonetheless saw themselves as falling at the bottom of the power structure with agency, planning, and nonprofit officials controlling the outcomes of the watershed management decisions.

Second, the implementation of the federal Endangered Species list has made the Salt Creek tiger beetle a central and controversial issue in the future of the conservation of the Eastern Nebraska Saline Wetlands. A number of the interviewees (including several agency professionals, nonprofit representatives, and landowners) expressed the belief that getting the Federal government involved in this process was not a good idea for the future of the salt marshes because it has made the issue a more contentious one. Third, natural resource agencies, planning, and nonprofit professionals believe the interagency partnership and the Endangered Species list have given them the capacity to better control the future outcomes of this area.

---

[13] Weber, E. P. 2000. "A New Vanguard for the Environment: Grass-Roots Ecosystem Management as a New Environmental Movement." *Society & Natural Resources* 13:237–259.

Other topics and minor themes include different perspectives and their impacts on planning for the future of the saline wetlands; a lack of trust between landowners and agency officials; and a general fear of the unknown by landowners.

Differences in Perceptions

The planners; local, state, and federal agencies; voluntary organizations; and the local landowners all have different opinions about the wetlands and their need for protection. Multiple levels of government have responsibility and power over decisions that affect the saline wetlands. Local public agencies have directives to plan for, design, and implement comprehensive land use planning and subdivision regulations, while federal agencies have national mandates to protect endangered species.

The Planners

During the interviews, it was apparent that the city, county, and long-range planners primarily see the importance of the Eastern Nebraska Saline Wetlands in terms of flood management for the city of Lincoln. Their role as planners gives them the responsibility of protecting the city and its inhabitants from flood damage. By protecting these saline wetlands from development, the planners feel that they are doing their job. As one planning official puts it:

> We do floodplain mapping primarily for our watershed master plans, to make sure that we use that to regulate ... not creating rising flood height ... so you're not getting an impact on the floodplain.[14]

The city of Lincoln has a Comprehensive Master Plan that maps out the possible areas of future growth and shows the possible green zones where they do not want development.

> We already had a comprehensive plan that talks about having new growth outside our floodplain, and that was reflected in the land use plan by showing environmental resources along the floodway, green space in the floodplain. And then some of the other floodplains were shown as agricultural chain corridor, and basically keeping those areas in open space and doing it right the first time in our new growth areas, and not encroaching into the floodplain and putting property at risk.[14]

According to one of the planning officials, the city planning department's goal is to "have a unified watershed plan for all the city of Lincoln and the future growth areas."[14] As another city planning official puts it,

> We use a land use map for going out 20 years in the future, so the land use map is very general in terms of urban residential commercial uses, etc. Most of the new growth areas,

---

[14]Lincoln 4 interview transcript, p. 4.

the flood plain is shown as future green space, and often in some cases ... that floodway is shown as environmental resources, and that's partly been shown as environmental resources to identify – it's a floodway; we're not looking for any development in it whatsoever.... We're not looking to see buildings in that area, but it could be incorporated as a part of the overall development.[15]

Clearly, from a land-planning perspective, the role of the city and county planning departments is to protect the public from future flood problems and to save green space for future use.

Ecological Protection and the Endangered Species List

Much like the planning officials, one of the goals of the local, state, and federal agencies and voluntary organizations is to protect the Eastern Nebraska Saline Wetlands. But instead of protecting these wetlands for the purpose of flood management, these agencies wish to protect the wetlands primarily for ecological reasons. Most of these organizations, such as the NRCS and NRD, are required to implement environmental protection laws passed by the federal and state governments. As one official from the NDEQ states,

We've got a lot of interest there [in the Eastern Nebraska Saline Wetlands] from an agency standpoint ... the eastern saline wetlands are priority under [section] 319 [of the Clean Water Act]. We do that not so much under the water quality part of it but restoration of aquatic habitat is one of the eligible categories, so that's the approach to the eastern saline wetlands.[16]

Additionally, a representative from The Nature Conservancy adds to this,

Our mission isn't to protect endangered species. Our mission is to protect biodiversity. If a landscape has an endangered species, we certainly factor that in, but that doesn't determine where we work.[17]

One tool that agencies can use to achieve natural resource and watershed management goals in the Eastern Nebraska Saline Wetlands is conservation easements on private property. In other words, the agencies offer to pay the landowner for the land as long as the governmental agency can control the practices that are acceptable on that particular piece of land. As a representative from the NRCS explains,

The landowner still retains full ownership of it, but we are paying them for the easement. We take the cropland, if it's cropland, or pastureland or whatever it is, and restore the wetlands back. So the saline wetlands, we're working with the Little Salt and in the Rock Creek basins, restoring some wetlands back in those areas. So with that we're doing a lot of activity.[18]

---

[15]Lincoln 1 interview transcript, p. 2.
[16]Lincoln 9 interview transcript, p. 8.
[17]Lincoln 7 interview transcript, pp. 12–13.
[18]Lincoln 6 interview transcript, p. 1.

Another official from the NDEQ adds that very few landowners who have land containing or surrounding the Eastern Nebraska Saline Wetlands are really interested in easements. According to the NDEQ official,

> In the Salt Creek system, nobody's really been interested in an easement.... I think we have 15 to 30 permanent easements. Most of them [the landowners] just say, "We're not interested in easements. We'll sell or we won't," flat out. I mean, it's there as a tool; it's just nobody's interested.... Well, I think it's a development issue. If they attach an easement to it, then it gets more complicated, because if you sell... the easement, that is a permanent easement. When the developer owns it, they would rather you just sell it off to a state agency or the City or whatever and then they don't have to deal with it. All they do is develop the upland and sell it to homeowners. Otherwise, they'd sell of the plots up above and they'd still have this 20 acres down there that they were responsible for. And they don't want to do that.[19]

Landowners

The landowners interviewed had some very different perspectives than those of the city planning and local, state, and federal agencies and voluntary organizations. Their outlook primarily involved their ability to control their own land management decisions and their economic well-being. In the words of one landowner, many other landowners in the area felt that they were powerless against the oncoming threat of losing their land to beetle habitat (L = landowner; I = interviewer)

L *So they proposed at one time a protected area that covered, I've forgotten how many acres – a hundred thousand acres or something that you couldn't develop or anything. So everything within two or three miles of any estuary going into Salt Valley you couldn't do anything with. And there was such an uproar over that. So you've seen these maps in the paper where every farmer would think – "Well, there goes my retirement income. I was planning on selling this, and I can't do it anymore." So there's a lot of fear over it.*

I *So do you think that's why so many acreages have went up?*

L *Yes, yup.*

I *Because of fear?*

L *Yeah. People were selling because they were fearful they wouldn't be able to in the future. So I know the 80 acres there and the 80 acres there, that was reason – because the owners of both properties told us.*

L *I had a bunch of people who ... talked to me ... they were terrified, and you would hear them say things like, "If any of those government surveyors come out on my property, I'm going to shoot 'em, because they're trespassers."*

I *Oh, no.*

L *Oh, yeah, and I don't think any of that happened, but people's feelings were really high.*

---

[19]Lincoln 9 interview transcript, p. 13.

Another landowner who has an easement on part of his property talks about the lack of control of land management decisions on his land,

> While we've done the easement here, we still own the property; but for land use, like having cattle out there or haying, I have to request permission and then send a separate contract for this year – you can go ahead and do it. But then they put a requirement on it that you can only do it in this window so that the grass will be X high for the fall for wildlife.[20]

Rather than being concerned with the biodiversity of the Eastern Nebraska Saline Wetlands, the landowners are primarily concerned with the economic value of their land and whether implementation of the Endangered Species list could infringe on their rights as property owners and as land managers.

Several of the landowners felt that the federal listing of the Salt Creek tiger beetle was not completely necessary. As one landowner states:

> As far as I'm concerned the whole thing, it's just blown out of proportion – because it's not a separate species, it's a subspecies. And the ecological significance of it, the significance of its limited gene pool is inconsequential to anything. It's just gotten blown out of proportion, and there really is a debate as to how many of them there are.[21]

The landowner goes on to add that:

L *It [the case of the tiger beetle] is being used now as an example of the excessive government regulation in Washington. And the people who were trying to delist the bald eagle, for example, are using that as an example that Fish and Wildlife has just gone crazy.*
I *So, you think that this particular beetle is an example of…?*
L *Right, as to how ridiculous it is.*[22]

Another landowner points out that although he has lived in the area all of his life, it was just recently that people in the area began hearing about the Salt Creek tiger beetle. The landowner states:

> Years and years ago, of course we had never heard of the tiger beetle … or I never or other people around here never even knew they were there. But nobody had ever heard of it, you know… I would say …. maybe about 12 years ago was when we first heard anything about them … it was probably 12 years … around 1997 or 98 when they started talking about them.[23]

Additionally, two representatives from The Nature Conservancy (TNC1 and TNC2) also point out that they are not sure that classifying the Salt Creek tiger beetle was the best move by the federal government. According to the interviewees,

TNC2 *The whole tiger beetle…. not sure that was the best strategy.*
I   *You mean the identification of the beetle?*

---

[20]Lincoln 11 interview transcript, p. 14.

[21]Lincoln 11 interview transcript, p. 5.

[22]Lincoln 11 interview transcript, p. 6.

[23]Lincoln 12 interview transcript, p. 3.

TNC2   *To list it on the federal ... because it really polarized the ... I mean, when we started, we were really hoping to keep it centered on the watershed, the ecosystem, the community things going on there. Now it's like, well, we can't develop this because of that stupid beetle. That is not where we wanted to be.*

TNC1   *And there may be some good things to come out of it, because it may attract research dollars and that kind of thing. It does give you some teeth if you really want to stop something from happening. But, man, it just changed the whole tenor of the discussion. So we were sort of disappointed that it actually finally got listed.*

TNC2   *The ESA [Endangered Species list] designation, I don't know who it helps, but especially you know Nebraska is 97% in private ownership. So when you start talking about an endangered species and impacting an endangered species, 97% of that is going to hit private lands, statistically speaking. And so you're really talking about infringing upon people's property rights, and they get pretty defensive about it. And I can understand that. And we've seen fights, just in the last five years, where it's gotten ugly on different species. You know, they talked about listing the prairie dog, and it brought everybody out – you know, why would you do that? The tiger beetle. And it does seem to make any discourse that much harder to have because then groups like ours are viewed as – you're here to protect the blankety-blank bug or the blankey-blank bird.*[17]

One of the landowners interviewed pointed out that, "a lot of landowners are very skeptical of environmentalists"[24] and then went on to add that, "the tiger beetle issue has become a divisive issue"[24] between landowners, developers, and those wanting to conserve the wetlands.

Prior to the implementation of the Endangered Species list, with the location of the Eastern Nebraska Saline Wetlands on the outskirts of Lincoln, a lot of landowners in the area looked at their land as retirement income for the future. As a city planning official points out,

> There was an effort that was led by The Nature Conservancy up in Little Salt Creek ... about conservation, etc. And they made the mistake of assuming that the farmer and landowners up there would want their land protected and not developed. And it was actually quite the contrary – the land was their nest egg, and they were assuming they were going to be able to sell it to a developer ... I mean, there are a couple, like there's a historic farm up here and some others where it's a long-time family holding and [they do not want] development at all. But there are certainly areas where, yeah...[25]

The Endangered Species list impacts the economic development value of the land these landowners wish to sell. Some of the landowners are selling the land as fast as they can so they will not have to deal with more regulations. As one landowner

---

[24]Lincoln 10 interview transcript, p. 1.

[25]Lincoln 4 interview transcript, pp. 11–12.

points out, there is a lot of nervousness among the landowners about the endangered species habitat:

> There is nervousness, and a big part of the problem is they [the landowners] really don't understand what the future holds. And a lot of that is just misinformation. I know that some of the early hearings on the wetlands here and everything, they weren't handled well. People went and misspoke, representing the state and feds, and they were just nasty. People would storm out of the meetings.[26]

Furthermore, one of the landowners interviewed currently has land containing saline wetlands in an easement. When asked about the landowner's neighbors' reactions to that easement, the landowner stated,

L *We have neighbors who are not happy that we did the easement. In fact, I had several ... talk to us, to me specifically after we had all the little white signs put up and they saw the surveyors out here repeatedly.*
I *Why would that be?*
L *Government control. They don't want government involvement in their farming operations at all.*
I *So, then with your easement, do they think that affects them?*
L *They're worried that it will. And then initially I think some of them were worried that the property was being designated as tiger beetle habitat – which wasn't the case, and in fact as I mentioned earlier, Fish and Wildlife came out to see if it was and said, no, it's not. They were worried that it was just the government reaching out and that they were going to be restricted with what they could do with their land. I actually had one neighbor who said that if they ever saw the surveyors out again, he was going to sue the federal government in court for trespassing.*
I *Even though it wasn't on the person's land?*
L *Right. Didn't matter. It was adjacent.*[27]

The landowners interviewed all perceived that landowners in the area are wary of government involvement in land management. Further, each landowner brought up the fact that the Endangered Species list has forced the landowners in the area to take measures to protect environmental quality. As one landowner put it, "*landowners take federal involvement very seriously because the fines can be horrendous.*"[24]

Interagency Partnerships

The SWCP was formed by local, state, and Federal agencies, The Nature Conservancy, landowners, community members, and planning officials during the time when the Endangered Species list first became an issue. Partnership members realized the

---

[26]Lincoln 11 interview transcript, p. 16.

[27]Lincoln 11 interview transcript, p. 15.

Endangered Species Act regulations might make this watershed a contentious one, so they acted to try to bring benefits to all involved. A representative of the Game and Parks Commission summarizes it best,

> One of the things we've tried hard within the partnership to emphasize … is that the partnership is about the whole system, not just the tiger beetle. And it's been hard to keep that distinction because people, once they hear "saline wetlands," because of the press coverage, everybody thinks tiger beetle – it's all about that bug. And it's like, well, that's part of the system, but our interest in saline wetlands is much broader. It's green space, it's the bird communities and other plants and diversity that these wetlands support. It's flood control, it's water quality improvement – other things that the wetlands can provide are also important and certainly should be important in an urbanizing environment like this.[28]

With regard to community collaboration and environmental protection, a number of planning and agency officials feel that the overall outcome of the partnership has been positive. One NRD official demonstrates this point:

> I think it's been a real good partnership. Sometimes you get into partnerships and particular parties maybe don't get along or they're adversarial, and I don't see it at all with this group. I mean, it's a pretty good group … there's some issues that you always have a few growing pains here and there. But everybody I think complements everybody pretty good with the partnership, so I think it's been working pretty well.[29]

Another agency official from the Game and Parks Commission adds:

> I spend most of my time with partnership-type activities and different government and private entities throughout the state on wetland conservation projects. And this is a good one. Certainly it's as valuable to sit down periodically and talk and get to know people and have the same people there all the time so you get to know the personalities, a level of trust develops so that you can work together, that there's not some hidden agenda that some-body's trying to promote, that you're open and honest and you work together on a set of common problems. And certainly the five entities in the partnership are doing that, so, yeah, it's been really positive.[30]

Conclusion

The location of the Eastern Nebraska Saline Wetlands on the outskirts of an out-ward-growing city creates land use management and environmental protection tensions that are not easily resolved. Since the wetlands are considered a part of Lincoln, but also include land once or currently owned by local landowners, resource management decisions in this watershed must integrate the diverse and

---

[28]Lincoln 8 interview transcript, p. 6.

[29]Lincoln 5 interview transcript, p. 11.

[30]Lincoln 8 interview transcript, p. 8.

often conflicting perspectives of rural/urban, city/county, local/state/federal, and private/public interests.

The Endangered Species list has made integrating these differing perspectives difficult because it often favors urban and public interest over private. However, through the collective actions of the Eastern Nebraska SWCP, representatives of the many sectors have been able to establish a dialog on how these unique environmental resources can be protected in a manner that is best for all involved.

Chapter 14
Cross-Cultural Collaboration for Riparian Restoration on Tribal Lands in Kansas

Charles J. Barden, Lillian Fisher, William M. Welton, and Ryan Dyer

Introduction

American Indian tribes are recognized as sovereign nations within the borders of the United States by the federal government. As such, they have the right and responsibility to set their own laws and standards regarding environmental quality within the boundaries of their reservations. This sovereignty sometimes puts them at odds with local, county, and state agencies who are similarly charged, but whose best interests or value systems may run counter to those of the tribe. Juxtapose this tension upon a long history of ethnic cleansing, broken treaties, forced acculturation, general ill treatment, and the sometimes well-intentioned, but ultimately harmful acts of government entities, and there is little need to wonder that tribal communities may treat offers of outside assistance with suspicion. On the side of the assisting entity, a land-grant university, for example, there is a cultural tendency toward a big brotherly approach of doing "what's best for the tribe" without recognizing that the tribe may have its own idea of what is best for them, or that this may differ from the land grant's plans.

Collaborative partnerships must recognize the sovereignty of tribal government, the dignity of tribal people, and the uniqueness of tribal culture to overcome distrust, and to accept and integrate tribal authority and decision mechanisms if they are to effectively address shared environmental concerns. This case study reports on lessons learned during a decade of cooperation between Kansas State University (KSU), Haskell Indian Nations University (Haskell), and the Prairie Band Potawatomi Nation (PBPN) on water quality education and riparian buffer adoption. Traditional tribal beliefs prioritize water as the most sacred of elements and provided the basis for engaging in cross-cultural collaboration on streambank stabilization, riparian restoration, and watershed bioassessment activities.

C.J. Barden (✉)
Horticulture, Forestry and Recreation Resources, Kansas State University,
2021 Throckmorton Manhattan, KS 66506, USA
e-mail: cbarden@ksu.edu

L.W. Morton and S.S. Brown (eds.), *Pathways for Getting to Better Water Quality: The Citizen Effect*, DOI 10.1007/978-1-4419-7282-8_14,
© Springer Science+Business Media, LLC 2011

The Prairie Band Potawatomi Nation

The PBPN reservation is located in northeast Kansas. The reservation is 11 miles by 11 miles, hence 121 square miles of rolling prairie, woodlands, crop, and pasture land, cut by numerous perennial streams. Land within the reservation has a checkerboard pattern of ownership, with more than 34,000 acres owned by the tribe, of which, some is held in trust by the Bureau of Indian Affairs, while much of the remainder is owned by non-Indians. The Potawatomi themselves are not historically a farming people and the croplands are largely leased to white farmers who use standard crop production practices relying on fertilizers and pesticides. The tribe maintains a herd of approximately 140 bison and limited vegetable gardens and orchards. Water for residents comes from the local rural water district, but the Potawatomi monitor the surface waters on the reservation because these are used for recreation and subsistence fishing. The reservation is located about an hour's drive from Haskell Indian Nations University, one of two intertribal colleges in the United States. Haskell's student body is comprised entirely of Native American students drawn from all 50 states and more than 130 different tribes. Haskell attempts to weave cultural awareness into the curriculum, including classes and research efforts.

The collaboration described here began with a Kansas State University Extension demonstration project entitled "Bioengineering and Riparian Buffers – Building Tribal Capacity to Improve Water Quality in Kansas." The USDA CSREES National Integrated Water Quality Program had established a priority opportunity for extension water quality outreach to underserved audiences. Dr. Charles Barden, primary author, initiated contacts and planning with the Native American tribes in Kansas. The specific history of conflicts and negotiation that established this successful cooperative project is discussed in the general context of challenges and opportunities for working with tribal organizations.[1]

First Contact: Overcoming Distrust

The environmental departments of the four Kansas tribes (PBPN, Sac & Fox, Iowa, and Kickapoo) were invited by Bill Welton (coauthor) to consider proposing a cooperative project. Welton, an environmental science professor at Haskell Indian Nations University, facilitated the initial discussion. The only tribe to accept the invitation was the PBPN. Tribal representatives attending the first meeting included the PBPN biologist and an interested tribal member. Tribal leaders were wary and uncertain about the idea. In an earlier "cooperative" effort – an organic orchard

---

[1] Barden, C. J. 2003. "Lessons Learned from Collaboration with the Potawatomi Nation." Presented at sixth International Union of Forestry Research Organizations (IUFRO) Extension Working Party Symposium, Troutdale, OR, September 30.

planting – a KSU faculty member had failed to follow through. Grant funding was lost, and the tribe lacked the resources to continue the work.

Once assured that KSU Extension forestry would not "walk away" after starting the project, the tribe was willing to discuss details of a new joint effort. While the tribal representatives only wanted to have tribal members hired for the tree planting and other fieldwork, they ultimately agreed to develop an integrated crew of three KSU students and three tribal members. The tribe selected its own crew members. KSU included the campus Native American Student Association in their recruiting. As a result of the meeting, a letter of support for the proposal from the tribal council was secured within a week. The proposal was funded late in 1998.

Appearing Before the Tribal Council

The project was a cultural as well as a scientific collaboration. As project plans were developed and implemented, the KSU partners found at every stage it was necessary to work with the Tribal Council and through Tribal departments rather than directly with individuals. Streambank stabilization practices were to be installed in spring 1999. To secure use of some of the tribe's heavy equipment, as proposed in the project, the KSU Principal Investigator had to appear before the Potawatomi Tribal Council, present the project plan for the riparian restoration, and request use of the tribes' heavy equipment.

The first scheduled meeting was canceled when the council got behind in their agenda. At the following meeting, the presentation seemed to be well received, until one of the council members asked pointedly, "Did you just *assume* that we would provide this equipment for your project?" Tribes are sensitive about their limited resources, and despite early discussions, the tribal leadership still felt no "ownership" of the project. When they were told that, with donation of their opera-tors and equipment, multiple streambank sites could be established, the tribal coun-cil unanimously approved the use of tribal equipment, although they left it up to the departments whether there would need to be a charge for its use. The departments never charged for use of their equipment or operators, on this first project.

The importance many Native Americans place on the natural world and relation-ships within it cannot be overemphasized.[2] Red elm (*Ulmus rubra*), a tree with declining populations due to its susceptibility to Dutch elm disease, is the preferred fuel for Potawatomi ceremonial fires. Black walnut (*Juglans nigra*) is a valuable species widely planted in Kansas, for both nuts and timber. Tribal leaders requested these tree species be included in the buffer plantings, so we added both species.

Overcoming the distrust with which the KSU partners were met during initial exchanges was achieved through a three pronged approach: a trusted third party,

---

[2]Deloria, Jr., Vine and Daniel Wildcat. 2001. *Power and Place: Indian Education in America.* Golden, CO: Fulcrum Resources.

Bill Welton of Haskell, was engaged to introduce them; an effort was made early to meet with and engage the tribal council in the project; and the tribe's suggestions were listened to and incorporated into the project.

Field Work Begins on the Reservation: Building Trust

The initial fieldwork – tree planting and installing revetments – was carried out in early spring in cold weather. Recruitment of students for outside work at this time of year is not necessarily easy. Ultimately three students from KSU College of Agriculture and two Potawatomi college students from Haskell formed the crew that did the heaviest work. Three Potawatomi high school students were hired to plant trees the following week.

One of the KSU students had worked the previous summer for an agricultural chemical application company in the area. Although he had applied fertilizer and other treatments to fields on the reservation, he remarked how he had never met or dealt with any tribal members. His Potawatomi co-worker told him that the tribe was originally hunters and warriors from the north woods (Michigan, Wisconsin, and Ontario), and they had little interest in being farmers. Indeed, most of the tribal agricultural land was leased out to non-Indian farm operators.

Conservation techniques demonstrated in this project included cedar revetments and willow posts to stabilize eroding stream banks, and the planting of native prairie grass and forest buffers to reduce pesticide and nutrient runoff into streams. Staff from several tribal departments, as well as student tribal members, worked alongside KSU faculty and students and other agency personnel to install the demonstration areas, thereby building tribal capacity to apply these conservation techniques.

The college-aged crew worked well together, performing physically demanding work, often in cold and wet conditions. A cedar revetment was installed, which involved dragging whole trees into position on the stream bank, and then anchoring them by driving stakes either with a sledgehammer or a gas-powered jackhammer. Each day we would come in from the field to eat lunch at the Potawatomi environmental office, next to the American Legion post. After several days, a tribal elder invited us to a tribal celebration on that Friday, to dedicate the opening of the first paved road on the reservation. He said "You are working to make our home a better place, please be our guest on Friday." When we mentioned to the tribal biologist how persistent the elder had been in asking us to attend the road celebration, he said, "Several tribal members have noticed the hard work of the crew, and they really want you to come to the event." When we suggested the elder might just be inviting us to be polite, the biologist said an elder would not likely do that, and certainly not twice!

After 4½ days of demanding outside work, we were able to attend a portion of the celebration. The project crew was indeed made to feel welcome, and the generous meal of buffalo burgers was a welcome respite from the sack lunches we had been eating all week. The celebration was an excellent cultural experience for the KSU students.

The following week, the tribal high school students worked hard planting the needed trees and shrubs in the riparian buffers. The plantings were mowed and weeded several times that first summer. The only setback occurred in November, when beaver moved into the revetment site and cut off many of the willow poles and stakes we had installed. When it was suggested that the beaver could be eliminated the tribe emphatically disagreed. The Potawatomi have a prophecy that states: "When the buffalo, beaver, and eagle return, the tribe will be prosperous again." The beaver could be trapped for pelts, but not just to remove them from an area. The willow cuttings were virtually wiped out, and did not resprout.

Only two complaints were heard about the initial project. One was from the county extension board, which questioned why KSU Extension was working so much with the tribe, when they do not pay local taxes that support the county extension agent. The criticism was easily deflected, because the project was supported through federal grant funds. The Bureau of Indian Affairs agent relayed that the other tribes were a bit jealous of the assistance that KSU was providing the Potawatomi. He simply reminded them that they chose not to attend the proposal-planning meeting they were invited to the previous summer.

During spring break in March 2000, a new crew of KSU college students was hired, along with tribal members, to install another cedar revetment to control streambank erosion, and to plant three more riparian buffers. To replace the beaver-eaten willows, sycamore seedlings were used, because beaver rarely feed on that species. In this year, the red elm was received from the nursery and added to the riparian buffers. While we were working on the site, a tribal member came by to see what we were doing. He was very interested in the red elm planting and we gave him four seedlings to plant. That fall we were checking the planting, and again he came by to visit. When asked how the red elm got through the dry summer, he replied he had not checked on the seedlings after he planted them. "If they make it, it was meant to be," he said.

The project employed one of the Potawatomi youth for the summer, to help maintain the four riparian buffer tree plantings and to help get the neglected fruit orchard back under management with irrigation, weed control, pruning, and fertilizing. The project had connected the tribe with another extension fruit expert who advised them on orchard management. This helped KSU "make amends" for dropping the orchard project several years earlier. The apple orchard has since become quite productive, and has been greatly expanded with the addition of grapes, brambles, cherries, and plums. The fresh fruit is primarily used at the early childhood and senior centers on the reservation.

Using a crew made up of both tribal and nontribal members was integral to our success. Working hard in a conspicuous place helped tribal members see that the KSU partners were sincere in the stated goals of our project. Attending the tribal celebration and helping out everyday people in little ways, whether it was providing a summer job to a teenager or giving a few extra seedlings to a tribal member with an interest in red elm, provided avenues for the Potawatomi to get to know KSU faculty and students better and to make it easier to work on future collaborations. Respecting their priorities and integrating them with our own, (i.e., giving preference to the beaver over the willow), showed that we respected tribal values.

Called Before the Kansas Legislature

In November of 1999, the KSU project leader was asked to appear before the Joint Legislative Committee on State/Tribal Relations in Topeka, to provide a presentation on the riparian restoration project. In the statehouse chambers, there was palpable tension between the state legislators and agency leaders on one side of the room, and the tribal members on the other side. The project leader was directed to sit on the "state" side of the room and many people were watching closely when he crossed the room from the speaker's area to greet the Potawatomi Chairwoman and chat with tribal members in attendance.

Our successful project was the only good news on the agenda that day; the only positive example of state–tribal relations. Wedged between heated testimony and arguments about the tribes selling state tax-exempt gasoline, the legality of tribe-issued license plates, and the spillover effects of tribal casino gambling, our slide show depicted excellent cooperation between a state-funded university, several government agencies, and the Potawatomi tribe. State governments have difficulty dealing with the federally granted status of tribes as "sovereign nations" within the state's boundaries. State agencies are not used to negotiating on taxes and fees, but we believe that if they would compromise on these issues with the tribes, the differences would not be insurmountable. Increased flexibility and compromise such as were developed for the riparian buffer project is an effective way to overcome some of the most problematic roadblocks.

Using the Demonstration Sites: Building Partnerships

In 2001, the Kansas Alliance for Wetlands and Streams (KAWS), an association of organizations and individuals interested in conservation, met at the Potawatomi tribal headquarters and toured the riparian buffer and streambank stabilization sites. Later, the tribal biologist agreed to serve a term on the KAWS board of directors. Classes from both Haskell Indian Nations University and KSU take field trips to the demonstration sites. Bureau of Indian Affairs land managers have also toured the sites. Research data on the effect of fabric mulch and tree shelters on seedling growth has been collected, and has proven very valuable.

As of 2003, the riparian buffers had become well established. The bur oak, pecan, green ash, and black walnut trees averaged over 8' (almost 3 m) in height. Most sycamores were over 15' (over 4 m) tall. The shrubs (plums and bush cherries) bore fruit for the first time that year. However, the red elm had been mislabeled from the nursery. The seedlings turned out to be winged elm (*Ulmus alata*), a shrubby, southern species that suffered winterkill injury and heavy deer browsing. The cedar revetments have stabilized almost a thousand feet (more than 300 m) of streambank, but some erosion still occurred where beaver have burrowed into the banks, and removed the streambank willows.

Even though the federal funding of the original project is long over, the tribe and the authors have continued to manage the sites, consult on forestry-related issues, and develop further collaborative projects. In 2006, KSU, Haskell, and the Potawatomi submitted a joint proposal for a research-demonstration project, which secured funding from the USDA Tribal College Research Grants Program. This second project was to document the long-term effect of the earlier cedar revetments, and to compare that with a control, untreated site, and a stream reach using rock weirs for stabilization. Haskell students learned field research techniques by collecting physical survey and bioassessment data. These same sites were used for the National Tribal College Research Workshop in May 2007. Findings of the bioassessment portion of the project were presented to the PBPN tribal council by (coauthor) Fisher in May 2008, along with scientific posters that were created by Haskell students involved in the research. The council was very receptive to the findings and was particularly pleased that Haskell students had benefited from participating in research on tribal lands.

A third collaborative project was begun in late 2008, again with funding from the USDA Tribal College Research Grants Program, to assess the streambank stability and riparian zone health of streams throughout the reservation. Walking in-stream surveys were conducted to collect geo-referenced data. Using the same strategy outlined in the original project, the survey teams were comprised of KSU students, Haskell students, and tribal members working in mixed groups of twos or threes. The PBPN Environmental Protection Agency (EPA) has been very supportive of this effort, providing us with up-to-date GIS maps of reservation land ownership boundaries, hiring tribal staff to help with this specific project, and providing a vehicle for transporting teams. The PBPN EPA staff also hosted the crew at an Indian taco lunch toward the end of our field work. This project will culminate with training on the installation of a streambank stabilization site with invitations extended to the three other Kansas tribes, the Kickapoo, Sac & Fox, and Iowa to attend the demonstration. The PBPN Department of Roads and Bridges will provide the heavy equipment and operators for the installation.

Another example of progress in the development of good relations between land grant universities and the PBPN is highlighted by the tribe's request for assistance with a problem unrelated to water quality. In 2008, the PBPN was having trouble with grazing management of their bison pasture. They turned to Barden, a forester, to find appropriate connections with other KSU faculty to develop a comprehensive herd and range management plan that would improve herd health and reduce the need for winter feeding. Barden served the intermediary function in this case, facilitating a meeting between the tribal council and KSU range specialists, and arranging a field trip to the Konza Prairie for tribal council members, tribal elders, and tribal staff. Konza Prairie is a long-term ecological research site, owned by The

Nature Conservancy, and managed by the KSU Biology department, where the tribe reviewed their bison management plan and practices. The partners are confident that this collaborative relationship will continue into the future.

Return of Red Elm

Correcting the mistake of planting winged elm instead of red elm in the first riparian buffer planting took several years. Because red elm was not available in the nursery trade, seed was collected from positively identified red elm and planted in a small nursery bed on the KSU campus, while others were grown in pots in the Kansas Forest Service greenhouse. The first red elms were transplanted into one of the buffer demonstration sites in spring 2007, and they are growing well. In April 2008, landscape-sized red elm saplings were ceremonially planted with the tribal council at the Potawatomi government center and each council member took a small red elm home for planting. One sapling was also planted at the childcare center.

Conclusion

Four riparian buffer and five streambank stabilization demonstration sites have now been established on the Potawatomi reservation, involving the planting of more than 4,000 trees and shrubs. The project has been successful on many levels, including the primary goal that the Potawatomi are now restoring other riparian sites on the reservation with the knowledge, skills, and contacts gained from the initial project. Also, the placement of both tribal and nontribal project employees in graduate education or professional employment in the environmental and agricultural fields has been rewarding. KSU students have gained a better understanding of the Native American perspective. The restored riparian demonstration sites are used annually for numerous field trips by many organizations.

The initial project has spawned three additional collaborative research projects between Haskell Indian Nations University and Kansas State University, two of which directly involved Potawatomi tribal personnel in carrying out the project deliverables. The authors have a heightened awareness of state and tribal conflicts, and increased sensitivity to local culture when proposing or conducting programs. We advise others wanting to work with Native American tribes to learn of their origins, their traditional resource uses, and be willing to incorporate things of value to the tribe in the project plans. Be ready to negotiate on points of conflict between the tribe and your organization. If possible, develop a good working relationship with one or two tribal staff or tribal council members who will be willing to champion your cause. One of the Haskell students (coauthor Dyer) who worked on the original cedar revetment back in 1999, was recently reelected to the Tribal Council position of treasurer. Having his continued support in the Council for our collaborations is invaluable.

Invest the time to make the relationship a long-term one. In general, short-term projects that leave the tribe with expensive-to-maintain equipment or procedures that require a high degree of training do little good and may, as in the case of the organic orchard that was left to the Potawatomi without adequate support, harm the chance for later collaborations.

As the KSU and Haskell partners learned more about the Potawatomi as a people, it became easier to include things in our grant proposals that are useful to them culturally, such as planting red elm trees instead of some other species, and to avoid offending them, such as suggesting we eliminate the sacred beavers that happened to be damaging some of the trees planted for our project. In turn, the relationship with the PBPN has matured and deepened. The tribe has grown from being primarily a recipient of an outreach project to being a true collaborator on these efforts to restore riparian areas and improve natural resource management on the reservation.

Acknowledgements Partial funding for these projects was provided by the USDA CSREES National Integrated Water Quality Program and the USDA Tribal College Research Grants Program.

Chapter 15
Getting to Performance-Based Outcomes at the Watershed Level

Lois Wright Morton and Jean McGuire

Performance-Based Environmental Management

Performance-based environmental management is the continuous assessment and adjustment of agricultural production management practices evaluated against impacts to the environment. This strategy is used by a few Iowa farmers to assess the effects of their practices on land and water resources and to identify opportunities for improving their financial bottom line. Today's conservation farmers are providing ecosystem services along with food, feed, fuel, and fiber products. Ecosystem services include clean water and air as well as plants and natural habitats that act as nutrient filters and support wildlife necessary for ecosystem balance. While farmers regularly track indicators of agricultural production (e.g., crop yields/acre, tons of milk/cow, and weight gain/day) and adjust their practices accordingly, they seldom track and evaluate on-farm and watershed-level indicators of the environment.

In Iowa's performance-based environmental management projects, agronomic decision tools such as models and in-field nutrient testing are being applied as quantitative and qualitative indicators of environmental impacts on soil quality, nutrient runoff, and soil conservation. As these indicators are tracked over seasons and years, the focus is on the question, "Are we doing better?" The key innovation in the performance approach is that it gives farmers direct feedback at the field, farm, and watershed level and an opportunity to discuss watershed progress with other farmers. "Doing better" is measured by first identifying intrinsic factors (e.g., tracking nutrient levels that have been scientifically proven to affect water and other environmental conditions) and extrinsic factors (e.g., tracking weather trends, market conditions, and what others are doing on the land in the watershed) to the farming process. These key factors are monitored on a regular basis to evaluate the impacts of current management practices on water and land resources.

L.W. Morton (✉)
Department of Sociology, Iowa State University, 317 East Hall Ames, IA 50011-1070, USA
e-mail: lwmorton@iastate.edu

L.W. Morton and S.S. Brown (eds.), *Pathways for Getting to Better Water Quality: The Citizen Effect*, DOI 10.1007/978-1-4419-7282-8_15,
© Springer Science+Business Media, LLC 2011

After 4 years of using performance-based management, the Hewitt Creek Watershed farmer-led group is seeing economic profitability and environmental goals achieved. Sixty-seven percent of the farmers in the watershed have participated in this program. With assistance from extension technical specialists, agronomic and environmental performance indexes have been tracked on 396 fields totaling 9,893 acres on 47 farms. Significant improvements in the Soil Conditioning Index (SCI) and Phosphorus Index (P index), reduced nitrogen use, and reduced sediment delivery to Hewitt Creek have been documented due to participants' management changes. The results have shown up in water monitoring data with a general trend of improved nutrient and suspended solids analyses and improved late summer dissolved oxygen levels. Improved diversity and quantity of macro-invertebrates were also found during semiannual evaluations.

How did this happen? How could a majority of farmers in a watershed become engaged in such an extensive effort to impact their lands and local streams? How did a group of farmers learn to track their nitrogen, phosphorous, soil condition, and soil losses? And what motivates them to continuously learn new and adjust current management practices to support better water quality? We attempt to answer these questions by presenting the concept of performance-based management by engaged citizens and its application to agriculture. The farmer-led Hewitt Creek Watershed performance-based environmental model is described using the farmers' own words, gathered through in-person interviews in 2005 and 2008. The farmers talk about how they first came to be aware of water problems and then involved in finding solutions. They discuss the assessment and goal setting process, making decisions to target specific concerns, measuring and evaluating the performance of their management practices, and continuously working at improving how they manage for economic and environmental outcomes. We conclude with the farmers' personal observations about changes in their own stream and their commitment to continue to champion this conservation systems approach.

Performance-Based Management

Performance assessment has been used for several decades by American industry to improve manufacturing processes and to meet safety and environmental regulations.[1] Many manufacturers discovered that they can follow all the rules but still produce an inferior product.[1] The goal of performance-based management is to reach and maintain excellence and effect continual improvement through constant assessment and adaptive management. This systematic "checking" on results leads to the discovery of places where an innovation or change could improve the production process.

---

[1] Wilson, Paul F. and Richard D. Pearson. 1995. *Performance-based Assessments: External, Internal, and Self-Assessment Tools for Total Quality Management*. Milwaukee, WI: ASQC Quality Press.

Continuous improvement programs like performance-based management are structured to provide timely, accurate, and constructive feedback so that production practices can be constantly upgraded with the final product meeting or exceeding quality control standards. Managing for continuous improvement assumes that the current norm for the process is inadequate and does not consistently, efficiently, and effectively produce an excellent product. This kind of management requires a shift in attitudes *away from a compliance mode*, which makes changes to meet minimum standards, *to a performance mode*, where continuous improvement is the main objective. Industry quickly learned that attaining performance-based management was based in day-to-day decisions and had to be everyone's responsibility. Individual and mutual objectives were more likely to be achieved with participation and involvement of all members of the production process. Wilson and Pearson[2] explain, "Each member of the team must measure and rate his or her own performance in relation to not only his or her own goals, but the group's goals and objectives."

Engaging Citizens to Practice Performance-Based Management

The idea of all members managing for improved performance has applications to citizen-farmer efforts to make better land management decisions throughout the watershed. Applied to watershed management in agricultural regions, "team members" are those who live on and work the land in the watershed. Few farmers have the specific technical expertise to address water quality issues on their own. However, their active engagement in watershed management is essential if shared local water problems, especially nonpoint source impairments, are to be effectively addressed. Further, each farmer must measure and rate their own performance to evaluate progress toward watershed-wide goals.

Epstein et al.[3] identify four practices that, if used properly and consistently, will lead to the achievement of measureable results on a local public issue. These practices are: (1) deliberately engage citizens, (2) identify indicators and measure results, (3) take action in response to measurements, and (4) community leaders govern for results. When natural resource experts offer opportunities, citizens can and do learn how to measure and monitor what is happening in their landscape and will work together to jointly undertake activities in their watershed to improve water quality outcomes. Helping farmers learn how to use new performance

[2] Wilson, Paul F. and Richard D. Pearson. 1995. *Performance-based Assessments: External, Internal, and Self-Assessment Tools for Total Quality Management*. Milwaukee, WI: ASQC Quality Press, p. 4.

[3] Epstein, Paul D., Paul M. Coates, and Lyle, D. Wray with David Swain. 2006. *Results that Matter: Improving Communities by Engaging Citizens, Measuring Performance and Getting Things Done*. San Francisco, CA: Jossey-Bass.

assessment tools increases their capacity to track and measure the results of their management practices. Giving them structured opportunities to talk about what these measurements mean for both their economic production and local water quality are motivators for changing practices and doing things that support individual and watershed goals.

Hewitt Creek Watershed

The Impairment

In 2002, the Iowa Department of Natural Resources (IDNR) identified a portion of Hickory Creek in the Hewitt Creek Watershed as "partially supporting" of aquatic life and listed it on Iowa's EPA Section 303(d) impaired waters list because it contained unusually high levels of nitrogen, phosphorus, and animal fecal coliform bacteria.[4] The implication of this listing was that farmers in the watershed faced the possibility that the IDNR would step in and regulate farming practices until the creek was delisted. This situation is increasingly common across the United States. It is estimated that nonpoint sources account for four billion tons of sediment, 80% of the nitrogen, 50% of the phosphorous, and 98% of the fecal coliform bacteria that pours into streams and rivers each year.[5] The US Geological Survey reported in June 2009 that spring nutrient delivery to the northern Gulf of Mexico was among the highest measured in 30 years.[6] Nutrients come from many sources, including fertilizers applied to agricultural fields, erosion of soils containing nutrients, suburban lawns, and golf courses.

The Hewitt Creek Watershed is part of the Maquoketa River Basin in northeast Iowa. Surface waters and groundwater feed directly into the Upper Mississippi River near Dubuque. This hilly 23,000-acre agricultural watershed is home to the baseball field made famous in the movie "The Field of Dreams." There are more than 80 farms in this region, most of which are livestock intensive, including several family-operated dairies. These farms have a potentially high environmental impact on water quality but are less likely to participate in existing conservation programs because of their small land base. Most operators who became involved in the Hewitt Creek Watershed group are young, aggressive, profitable managers.

---

[4]Iowa Department of Natural Resources. 2002. (http://programs.iowadnr.gov/adbnet/segment. aspx?sid=110).

[5]Novotny, V. and G. Chesters. 1981. *Handbook of Nonpoint Pollution.* New York: Van Nostrand Reinhold Company.

[6]U.S. Geological Survey (USGS). 2009. (http://www.usgs.gov/newsroom/article .asp?ID=2240&from=rss_home).

The Hewitt Creek Watershed Model: A Performance-Based Management Solution

A team of extension faculty and field specialists from Iowa State University (ISU) developed a performance-based environmental management intervention model that links farmer-led group interactions with agronomic tools as management feedback. A USDA CSREES (which became the National Institute of Food & Agriculture [NIFA] in 2009) National Integrated Water Quality Program grant, "Educational Program to Increase Citizens' Responsibility for Management of Agricultural Watersheds," received in 2004, provided funding to pilot test the model. Impaired watersheds in the Maquoketa Watershed Basin were offered the opportunity to be a pilot site. Several Hewitt Creek Watershed farmers volunteered to help organize a group in their watershed when they learned it was listed on the EPA 303(d) impaired water list. This group of farmers experimented with the team model (Fig. 15.1), creating The Hewitt Creek Model of performance-based environmental management.

We use the Model to document the performance-based management process and how it enabled farmers to work together to learn new management skills that reduce off-farm nutrient and sediment loss. Data sources include monthly, quarterly, and annual reports to funders as well as audiotaped and transcribed in-person interviews with seven farmers and two watershed technical specialists in 2005, followed by 12 interviews in 2008 with the original participants and three additional farmers who joined the project after 2005.

Awareness

The first step in the model is becoming aware that there is a problem. Awareness can be initiated by a community leader, an environmentally conscious farmer, an extension educator, or an agency technical specialist. Typically four to six people who live in the watershed are engaged first in one-to-one conversations and then

Fig. 15.1 The Hewitt Creek Model for performance-based farm and watershed environmental management

convened as a group to talk about water issues and options for solving them. This initial group becomes the catalyst for change in their watershed by inviting others to participate. One of the founding farmers of Hewitt Creek Watershed group says:

> I would say ... there's a good hard core group of 20–25 people.... People that are interested in making some changes and know there's changes that need to be made. The waterway being flagged as something that needs attention definitely drew their attention and they wanted to be part of something to clean it up and they know that they can do it as well as the government coming in and telling them how to do it, that it may be more costly to them.[7]

A swine and row crop producer in the watershed comments:

> ...nobody really was aware of how ... this water was sensitive to what we were doing. We were looking for maximum row crop production from fencerow to fencerow.[8]

In many water quality situations, the problem is externally identified and not "owned" by local residents. Before people will actively work at solving a problem, they must first believe there is a problem and that they have some capacity to resolve it. In the case of Hewitt Creek Watershed, the IDNR provided the information that there was a problem. The external identification of the problem was aggravating to local watershed residents and became an organizing motivator (e.g., to "prove" IDNR wrong). Part of their local frustration was not knowing how the "impaired" designation was arrived at. The first meeting of farmers convened by the extension agronomist field specialists[9] discussed the 303(d) impairment and how they could verify whether it was true or not. Extension arranged for scientists in the IDNR monitoring program to come and discuss their results directly with the farmer group. As a result, the watershed group allocated resources to do their own monitoring. Although they subsequently discovered that their results were congruent with IDNR, it was necessary that they come to this conclusion themselves. "Effective resource management relies upon having and understanding scientific data"[10] and the process of gathering their own monitoring data provided the opportunity to learn how to read and interpret the indicators of water quality.

---

[7]Data from the Hewitt Creek Watershed group farmer interviews were collected in 2005 by Annette Bitto and 2008 by Jean McGuire. Transcripts were audiorecorded, transcribed, coded, reconciled, and analyzed by the research team Lois Wright Morton, Annette Bitto, and Jean McGuire. Farmer and extension staff interviews are identified by number and date. This quote is from Farmer #2 who was interviewed in 2005.

[8]Farmer #4, 2005.

[9]One key to the success of this project was the commitment and knowledge the ISU Extension agronomists John Rodecap and Chad Ingels. Rather than assuming "expert" roles, they were willing to experiment and work alongside the farmers, letting local knowledge and readiness to learn guide when to share their scientific and technological information with the group and one-on-one.

[10]Dennison, William C., Todd R. Lookingbill, Tim J. B. Carruthers, Jane M. Hawkey, and Shawn L. Carter. 2007. "An Eye-opening Approach to Developing and Communicating Integrated Environmental Assessments." *Frontiers in Ecology and the Environment* 5(6):307–314, p. 307.

Assessment

During the group formation process, members become more aware of their water conditions, nutrients that could be contributing to potential problems, and assessment tools they can use to monitor their fields and the overall watershed. They begin to take baseline readings on their own farms to assess their current management practices. A farmer who joined the group in 2005 talks about what he learned from his assessments:

> …we had everything calibrated … I couldn't believe how much manure we were actually hauling on some of these places … now we're at the point where we have everything tested and we're hauling manure on places where it is needed.[11]

Another farmer lists some of his assessment practices:

> I've done some late spring nitrogen testing, and we will do a cornstalk testing this fall … I've done manure analysis … we've got to do a manure spreader calibration this fall and I just filled out that paper for that P Index thing.[8]

After a first year of assessment, farmers learn how useful the information is for making management decisions and engage in continuous assessment, which leads to continuous improvement in their management practices. Assessment includes not just soil tests and plant analysis but also seasonal observation of field conditions. One farmer notes that some of his waterways performed better than others and plans to make changes based on what he's learned:

> …another thing was waterways. We didn't have our waterways big enough. Especially after this year with all the rain, we've definitely seen the waterways we put in were a great big plus.[11]

It is in the baseline assessment step that the group discovers the actual environmental condition of their watershed by seeking research-based data. Assessment includes determining chemical and physical characteristics of the water, biotic indexes (fish, clams, and invertebrates); land use adjacent to water bodies; land use within the watershed; natural land and water habitat; and wildlife counts. The intent of the assessment step is to combine farmer local knowledge with scientific and technical information. This information provides a foundation for the group to identify performance indicators that they can easily use as feedback tools when they move to targeting and performance steps.

The extension field agronomists used this participatory discovery process to educate and develop consensus that there was a serious local water quality issue. The state's testing that established the pollution problem was used as a baseline for the project. However, the group wanted to do their own assessment. With assistance from a local college, they created a plan that identified what water tests would be conducted, where within the creek samples would be collected, and when the testing would be done. This effort established a systematic monitoring process to build a database of past and current trends. These data were used later to evaluate whether

---

[11] Farmer #9, 2008.

performance goals were met and to refine activities and practices to achieve goals. The Hewitt Creek Watershed group also decided who would collect the information, where it would be archived, and how the information would be shared and discussed publicly.

Goals and Plans

The next step in the process is for the group to set individual field, farm, and watershed-wide goals. These goals follow naturally from the assessment data that are gathered and shared by the group. The reluctance to reveal personal farm-level data is overcome by the extension agronomist assigning each field a code number that only the field owner/manager recognizes. Then the assessment data can be aggregated so the big picture of the watershed condition and overall management practices emerge. This step is essential if the group is to move beyond their individual management to thinking about watershed-wide management.

Underlying the application of performance-based management is the assumption that there is "agreement on what performance is, how to measure it, and how to use the results to devise effective means for performance."[12] Two kinds of goals were set by the Hewitt Creek group: watershed-wide goals set by the group and individual farm goals. Individual goals varied by farm operation characteristics, soil, and typography and personal goals for the business. However, participating producers encouraged each other in not just setting economic production goals but also conservation practices in support of water quality goals. Farmers reviewed the information collected in the assessment phase to determine how to best achieve his/her environmental goals while maintaining the profitability of his/her operation. These goals included improved soil condition (organic matter, water management, and soil carbon), reduced nitrogen or phosphorus delivery to the water body, and slowing down water flow, thereby reducing erosion and sediment loss.

Watershed-wide goals included working to remove the Hickory Creek segment of Hewitt Creek from the Iowa impaired waters list, improving biological water quality indexes and wildlife diversity, reducing pollutants, reducing stream flow velocity, as well as reaching community goals such as expanding recreational use of the water resources within the watershed.

Well, the long-term goal is to get it [Hewitt Creek] off the map, get it to where we're off the DNR's radar, whoever is watching this water quality thing. And if we could ... get it cleaned up, I think we'd be the better men...[13]

---

[12]Wilson, Paul F. and Richard D. Pearson. 1995. *Performance-based Assessments: External, Internal, and Self-Assessment Tools for Total Quality Management.* Milwaukee, WI: ASQC Quality Press, p. 6.
[13]Farmer #3, 2005.

> My vision would be every farmer in the Hewitt Creek watershed doing a practice ... at least one practice on every farm ... if every farmer would just try something to improve.[14]

Often personal goals and watershed-wide goals merge. One farmer reports three goals he lists when other farmers ask him why he participates. He tells them:

> ...we're trying to be better farmers at using our energy and our P and K. I tell the guys, you are livestock farmers. Either you better get on the bandwagon and figure out what you [need to] learn, or the DNR is gonna come in and set it up for you. Now which one would you rather have – be mandated or have a chance to write your own destiny?[15]

To achieve these goals, the group sought funding to cover costs of diagnostic testing and implementing the practices they selected. One member of the newly formed group was very active in the Farm Bureau and he, with the support of others, sought and obtained funding for the project from the local and eventually the state Farm Bureau organization. After the group had developed a cohesive structure and common goals, the leaders applied for tax-exempt group status, which enabled them to obtain additional funding from public sector programs, such as the Iowa Watershed Improvement Review Board (WIRB). With assistance from Iowa State University (ISU) Extension specialists, the Hewitt Creek Watershed group obtained additional funds and in-kind support for staff, demonstrations, surveys, and water monitoring from EPA Region VII, the USDA CSREES (NIFA) National Integrated Water Quality Program, ISU, the Iowa Department of Agriculture and Land Stewardship, Land Contractors of Iowa, the Iowa Corn Promotion Board, Upper Iowa University, and high school vocational agriculture programs in watershed counties. This variety of resources demonstrated to the group that there was support at the federal, state, and local levels available to assist their efforts to clean up their watershed.

Targeting

After their goals were set, the Hewitt Creek Watershed group met to set targets and prioritize management practices and activities to help meet their performance goals. They recognized that the priority practices needed to be acceptable and practical. The group then designed a locally managed incentive program to encourage participating farmers to implement practices that could be objectively measured and were most likely to improve water quality. The incentive program was unique in being designed to reward performance measures rather than the practices directly. In the first year, all participants received incentives to test and calculate the baseline performance index values on their farms. In subsequent years, they had the flexibility to implement, or test on a preliminary small scale, whichever targeted soil and nutrient management practices were suited to their current operation and within

---

[14]Farmer #7, 2005.
[15]Farmer #4, 2008.

their management and financial ability. They received incentives when their whole-farm index scores improved. The Dubuque County Farm Bureau provided $500 and the Iowa Farm Bureau Federation provided the Hewitt Creek group with a grant for $30,000 to pilot test their incentive program.

One dairy farmer used the incentive program to try fall cover crops to improve his Soil Conditioning Index score and reduce erosion on his row crop land.

It's like the first year I did it – my agronomist from the co-op was driving by. I [had] just got done chopping the corn [for silage]. I already spread the manure on it, and I was disking it and pulling the harrow, you know, disking it real lightly, pulling the harrow, just incorporating the manure and leveling it up. And he stops out in the field with his pickup, and he says, "What in the hell are you doing? You're the only guy in Dubuque County disking." I says, "Stop by tomorrow. I'll be seeding oats and rye." And he just shook his head, and I said, "Then stop back in about two, three weeks and take a look at it." And he's the same guy that weighs my corn, and he just can't believe it. Yeah, he's convinced now. He just can't figure out how we get that much corn out of such little nitrogen. He even said it this fall again – just didn't know how we did that.[16]

Other farmers discuss targeting nitrogen use through testing:

...the local fertilizer guy, he'd try to tell you we need P and K this year. I says, 'Really? Let's grid sample it first'.... A lot of these grounds we don't need anything on ... it costs money to do the grid sample, but ... we have the facts and figures ... we grid sample again ... see how much it changed [and then we know how much we need].[15]

We were always putting 150 units on. We're down to 75 now on the cornstalks and we don't put any on the sod ground and we can still grow 230 bushel of corn.[17]

Table 15.1 illustrates some of the activities and practices the Hewitt Creek Watershed group farmer cooperators undertook in their first crop year, 2005.

Performance

Data from the assessment phase combined with goal setting and targeting leads to changes in management practices. As farmers identify what needs to change, they adapt their management to get better environmental and production performance. The Hewitt Creek Watershed group farmers learned to use and evaluate the results of three cropland agronomic tools that measured nitrogen sufficiency/excess for corn production (late season stalk nitrate test), potential phosphorous loading from cropland (P Index), and improved soil quality (Soil Conditioning Index [SCI] to track increase in organic matter) on their farms. These tools are research-based tests and models currently used by extension, federal, and private sector specialists as part of Best Management Practices (BMPs) for crop nutrient and soil conservation planning. Reassessing the measures from year to year as qualitative indexes of environmental progress is an innovation that has made the performance

---

[16]Farmer #8, 2008.

[17]Farmer #5, 2008.

Table 15.1. Example of Crop Year 2005 Incentives for On-farm Activities and Practices

Number of Cooperators	Incentive Payment	Water quality improvement activity
14	$80	Two cornstalk residual nitrogen to compare 2 commercial N and/or manure rates. (Sample is 15 8-inch segments.)
5	$25	For each additional cornstalk test to refine N (limit $100 per operator, includes $15/sample lab fee).
3	$50	P soil testing and ISU interpretation to identify fields testing VH (more than 21 ppm).
7	$60	Manure applicator calibration to determine rate per acre.
14	$50	Manure analysis (testing) to determine available N, P & K.
13	$80	Complete P index on two fields to determine the risk of P loss—will also receive the Soil Conditioning Index (an indication of trend of soil organic matter management).
4	$200	Tillage alternatives or no-till field scale comparison of conventional practices demonstration with yield results.
6	$200	Field scale or small plot comparison of N or P rates with or without manure yields determined (ISU Extension will assist).
12	$300	Grid sampling 40 or more acres/operator.
25	$400	New grass waterways/per operator.
5	$200	Seed headlands or other buffers including along streams/per operator.
7	$10/acre	Cover crop seeded after corn silage harvest up to 40 acres/operator.
4	$120	Tall grass filter below feedlot.
4	$250	Earthen diversion or roof gutters to keep water off livestock lots. Catch basin to collect solids below feedlot.
5	$50	Self assessment of farmstead including livestock operation as appropriate.

concept workable at the farm and field level. Tracking their index results from year to year provides farmers the information they need to set additional goals and adapt their plans to better manage loss of sediment and nutrients into local water bodies. As one farmer points out, some of the new tools were easier to use than others.

> So when we started … it was a little hard to understand the Soil Conditioning Index and the P Index … But the other one, stalk nitrate test, that was…. We were wasting a lot on fertilizer – we learned that very, very quick…[15]

The incentive program that the Hewitt Creek Watershed farmer group designed focused on measurable performance outcomes. For example, they paid themselves if the weighted whole farm P Index was less than a phosphorous loss risk of 3 (2–5 is medium risk); and gave a bonus if the P Index was 2 or less or if all field tests were within or less than the optimum P university soil test range. They rewarded Soil Conditioning Index readings for each 0.1 increment above zero and identified practices such as cover crops and reduced tillage, contouring, and waterways as strategies for accomplishing the performance goal. Successful nitrogen performance management was measured by farm-weighted average stalk nitrate analyses that did not exceed 1,700 ppm. They gave a bonus if weighted averages of all analyses were less than 1,300 ppm and for wetland impoundment or drainage tile management put in place to reduce nitrogen.

While farmers were collecting information specific to their own fields using this process, the Iowa State University Extension field agronomist aggregated the information in a spreadsheet so the information could be anonymously shared with the entire farmer group. This sheet became a valuable education tool, helping farmers to better understand what each index measured and what practices would help shift the index numbers in a direction that might improve water quality. This exercise allowed farmers to evaluate how successful the group had been in reducing potential agricultural pollution to the creek. Further, this information helped the group to assess whether the performance-based management practices were helping them achieve their goals.

> You know, we don't know what's going on on the other side of the watershed because we don't have time to drive around the block all the time. But if they already have the data in front of you and they can tell you what they did for their tillage and their nitrogen and how many pounds of seed they put on and what weed control they put on and the stalk…, how much nitrogen they had in their stalks, and then the total yield they had on the end – well, you can look at that whole picture of that farm. And times that by all the farms that are on that sheet and then think – well, I can cut corners here or I could do something different here, which will save me a lot of money.[17]

One unique performance measure the farmers chose was whether new, nonparticipating farmers joined the group effort. When farmer participation significantly increased, the group paid each participating member a small bonus. The watershed group collectively decided on this bonus because they felt watershed-wide participation was the key to achieving their goals. One of the original farmers in the project said that seeing made believers out of some of the skeptics.

> Well, I'm participating, and I can see where this is helping, maybe that'll spark another person down the road to jump in … participate. That's the hope we've got, is by us participating that maybe down the road where these people, their neighbors can see an advantage and they may want to jump in and also help.[15]

A row crop farmer pointed out that the benefits of being a part of the watershed group were so substantial that he wondered why some farmers had not joined the effort.

> If a guy don't want to do it, that's fine, but you get everybody else around him doing it, then all of a sudden he sits there and says, 'Well, I'm not in it.' Well, I said, 'We've got enough cost figures now, we can come sit on your farm and say, hey, what do you want to look at? Do you want to look at 55 working units that are showing reduced costs savings? Why wouldn't you want to look at it?'[18]

---

[18]Farmer #3, 2008.

Evaluation

The cycle comes full circle with the evaluation of performance measures against individual and group watershed goals. Continuous, systematic monitoring of performance indexes and tests allow farm managers to adjust management practices for continued improvement. As farmers regularly used performance-based measures, they gained confidence in making future management decisions based on their findings. Unusually heavy rains and flooding in the summer of 2008 caused a great deal of erosion in eastern Iowa, but those farmers who had recently installed waterways saw the benefits.

> You know, it was just phenomenal the soil erosion this year. And this year we really appreciated the waterways that we put in the last two previous years. I mean, it was very much well worth our time to put them in. And this program didn't pay for the whole cost of putting the waterways in, but for what it cost us out of our own pocket, we more than got paid just for the minimum of soil loss that was caught in the waterways that now next spring we're going to take back and we're going to push back up on our ground again.[15]

Because a variety of techniques and practices were tried on different farms, all farmers in the watershed were able to learn from each other what worked and what did not. They developed BMPs and performance measures that worked effectively on the farms in their watershed. These practices and methods of performance measurement have become known and accepted in their community. Three farmers shared these thoughts about the project.

> And once you had [a] ... larger core of individuals that had similar results, well, how far can we push this thing.... so we get to the point where we had a train wreck, went too far, now we back up a little bit and sit down.... What's behind these things, a meeting after a meeting – we thought we ought to try next year.... You know, you push this, you push that. What'd you think of that? Well, this was good. Rye grass, that was all right, we seen some results from it, but we changed. And it's just like, and nitrogen anymore is one of the key issues you look at.[18]

> The nice thing about it – other farmers were doing some other things, so you don't have to be the guinea pig on everything by yourself. And then so everybody did something a little different, and all this information was gathered. And it was nice to see in the end some things that did work, some things that didn't work, and what worked one year and didn't work as good the next year. So now we can kind of try to curtail it a little bit different for the following year.[15]

> One farmer is going to believe the next farmer, number one, because he lived there, versus somebody that comes out from Iowa State that you don't know. Now, if you go down to my neighbor and try and teach him something, he's not gonna believe you. If I go down and tell him and he saw it on my farm last year, he'll believe me.[16]

The Effects of Performance-Based Management Process

The performance incentive payments set by the watershed group to promote environmental improvement were modest by the standards of current public conservation programs. Yet by 2008 it was clear to the participants of the Hewitt Creek Watershed group that their incentives were of less value than the knowledge they had gained over the past few years.

> If you looked at the group ... that got into this program in the beginning, they weren't in it for the money. They're in it to learn and if they can improve their operation, that's why they did it. I think the whole key behind this thing was to know what our indexes were. And the management decision behind that was to see where we could tweak things to make money. That's where we made our money at – it was the knowledge, not in what these guys paid us.[18]

Several farmers learned to keep and use farm management records, including a row crop farmer who helped get the project started.

> We are keeping more records probably. The biggest thing would be the manure sampling. We try to get that done with the stalk nitrate sampling. Keeping the stalk nitrate sampling probably has been one of the strong ones for me so that I can keep my nitrogen to where I want it and still get the crop I'm after. But that's something I never had before that I keep now.[19]

Some of the practices spread outside the watershed. One dairy farmer who was among the first to join the Hewitt Creek group has moved performance management to the watershed next door.

> My dad, he's not in this watershed and my brother-in-law isn't, my brother, and they all farm. And they're thinking different because of what we found out and from what I did and what other people in this here watershed did. So they're doing their management practices different also.[15]

In 2007, the Iowa State University Extension specialists surveyed Hewitt Creek performance incentive program participants. All reported they believed that the performance-based management process offered an effective conservation systems approach to improving water quality. Ninety-four percent of survey respondents said this approach encourages production and environmental management changes and 100% said the program has had a positive effect on the environment. All survey respondents reported that the performance program made their farming operations more profitable. Increased awareness of watershed and stream quality prompted 56 of 84 farm operators and owners to become project participants. In addition to peer pressure that led to second- and third-year new enrollments, many farmers thought there was a spillover effect on those who chose not to enroll. Neighbor-to-neighbor exchange of information was identified in the preproject survey as the most important source of resident information and was used to increase participation.

Even though the incentive funding for the project ended December 31, 2008, the watershed group continues to meet and is working on getting additional funding. Monthly meetings regularly draw 12–25 watershed farmer cooperators. The agenda always begins by reviewing the most recent water monitoring data. The group then proceeds to evaluating project progress, and fine-tuning the incentives to increase adoption of water quality management and performance improvement in their watershed.

---

[19] Farmer #2, 2008.

A Continuous Process

The Hewitt Creek performance-based management project has offered farmers learning opportunities to merge science-based protocols with deliberate and regular neighbor-to-neighbor discussions to give them new management skills and ways to assess whether they are "doing better." Central to their success is the setting of individual farm-level goals and group watershed-level goals with commitments and actions to implement practices that meet both. Like many industries, this farmer group applied a systems approach, using models, experimental trials, and observations of outcomes of neighboring farmers' practices to guide their management decisions.

> Ninety percent of them aren't there for the money, I don't think. That's not a big thing. It's a nice little incentive, like the $200 for the manure spreader calibration, yeah, that made it worth the while. Yeah, it's like, well, hey, this is worth my time. It's $200 – I can do that. But, no. As far as just being it for the incentive pay, no. The big incentive pay is … on the phosphorous, what we're seeing on commercial fertilizer in what we're saving by raising that Soil Conditioning Index score, on less tillage, what's that saving us, by our fuel man not pulling in.[20]

This watershed has not yet been removed from EPA's 303(d) impaired water body list. However, the group's members are getting better in managing for both economic viability and environmental sustainability. These farmers have learned an important principle of performance management: *incremental improvements happen when you practice day-to-day adaptive management that responds to systematic monitoring and assessment of the production process.* Some adaptations will be very successful, others will not be. Industry applications of the performance-based assessment process often report that any new innovation toward improvement is likely to have some embedded flaws.[21] When changed management practices do not yield the expected results, industry and the farmer practicing performance-based management will revise their plans, using what they have learned to fine-tune their production system. This will always be a continuous process because farmers must manage under dynamic and continuously changing conditions (e.g., weather, markets, labor, equipment, etc.).

The Hewitt Creek Model is a continuous cycle that builds on shared information and joint planning for getting to better water quality outcomes. Currently three other watersheds in northeast Iowa are using the same model to address similar water quality issues. In 2005, an early member of the project shared his hopes for the watershed group.

> There's fish in the creeks now, but to get a certain kind of fish that are good to catch and eat, that would be the top goal, I guess.[22]

---

[20] Farmer #7, 2008.
[21] Wilson and Pearson 1995.
[22] Farmer #6, 2005.

By 2008, one cooperator was pleased to report that nature was demonstrating that water quality in Hickory Creek had improved.

> I grew up around here and I said we used to go fishing in that stream on one of the farms I rent, and we haven't fished in there for years, years and years. And this past year one of the neighbors … was out with his grandson several times through the summer. I'd be coming across the road and he'd be sitting on the bridge with their Gator and fishing poles hanging off the side of the bridge. And I did stop the one time, and I visited with him a little bit. I said, 'Did you catch anything?' Yeah, about three or four catfish in the bucket, and I was real surprised to see that.

> And a neighbor to the south of me here…. The stream runs right by the house, and he's talking about the eagles that were around last year. He said, 'We used to see eagles out here when I grew up as a kid, all the time. We haven't seen them in years.' And he said this past year they were around that stream all winter long.[23]

The farmers of the Hewitt Creek Watershed have shown that it is possible for farmers to use performance-based management methods to achieve progressive improvement in environmental performance of their farm operations, to protect water quality and also maintain or improve their profitability.

---

[23] Farmer #2, 2008.

Chapter 16
A Farmer Learning Circle: The Sugar Creek Partners, Ohio

Mark R. Weaver, Richard H. Moore, and Jason Shaw Parker

Grassroots Participation

Those looking to implement grassroots participation in collaborative watershed management must be aware of key differences in the problems found in different watersheds. A "one size fits all" management model can never account for all the significant social and biophysical variables that determine success or failure. In particular, grassroots watershed organizations that consist of stakeholders who are the "consumers" of natural resources, such as farmers, pose different social and land use issues than are typical of watershed partnerships, or collaborations among existing public and private groups such as environmentalists, business interests, and government officials. In the case of partnerships among existing groups, the primary task is to find common ground or room for compromise among the different perspectives and interests that each group brings to a common resource, the watershed. In contrast, in the formation of new grassroots groups, the main task is to foster the development of a common perspective or a common set of beliefs regarding the watershed that will unify the group and guide future collaborative action.

We present a case study of the formation of a farmer-led grassroots watershed group, the Sugar Creek Partners. We analyze the formation process and outcomes using a modified version of a social science model that is used to explain policy change and learning at both national and local levels.[1] This framework, as we have

---

[1]The framework used here is a simplified version of "action core beliefs" as presented in the advocacy coalition framework. Sabatier, Paul A. and Hank C. Jenkins-Smith (eds). 1993. *Policy Change and Policy Learning: An Advocacy Coalition Approach.* Boulder: Westview Press; Sabatier, Paul A. and Hank C. Jenkins-Smith. 1999. The Advocacy Coalition Framework: An Assessment. Pp. 117–166 in *Theories of the Policy Process* edited by Paul A. Sabatier. Boulder: Westview Press; Sabatier, Paul A., Will Focht, Mark Lubell, Zev Trachtenberg, Arnold Vedlitz, and Marty Matlock. 2005. *Swimming Upstream: Collaborative Approaches to Watershed Management.* Cambridge: The MIT Press.

M.R. Weaver (✉)
Department of Political Science, College of Wooster, 130 Kauke Hall,
400 E University Street, Wooster, OH 44691, USA
e-mail: mweaver@wooster.edu

L.W. Morton and S.S. Brown (eds.), *Pathways for Getting to Better Water Quality:*
The Citizen Effect, DOI 10.1007/978-1-4419-7282-8_16,
© Springer Science+Business Media, LLC 2011

reformulated it, emphasizes six key components of the belief structure of an action-oriented group: (1) organization and process, (2) problem identification and definition, (3) causation and responsibility, (4) participation, (5) actions and solutions, and (6) values and goals. Each component is addressed in terms of a central question that organizers typically confront when they attempt to form a new grassroots watershed group.

The first and last components of the model raise critical value-laden questions: What kind of watershed group is appropriate given the circumstances? And, what is the basic vision or set of value commitments that will guide the group? The answers to these two questions largely determine the nature of the group that is formed and illustrate that organizers will necessarily confront normative questions requiring them to address the core values and beliefs of stakeholders. Organizers must also answer four additional questions that will force them to effectively communicate complex empirical information to stakeholders: What are the main problems that affect water quality and ecological sustainability in the watershed? What are the causes of these problems, and what individuals or groups bear primary responsibility for the impairments that exist? Whose participation must be sought to solve the problems? What types of actions or solutions need to be pursued by the watershed group?

We conducted a case study analysis of the Sugar Creek Partners utilizing multiple forms of data collection. First, we developed and administered a survey of residents who own property along the Upper Sugar Creek to determine existing beliefs about the condition and uses of the stream. Second, we engaged in participant observation of the meetings of the Partners. Third, we conducted in-depth, more open-ended interviews of selected residents of the watershed. For example, in the summer of 2005, a research team collected and analyzed oral histories of 30 residents living in the Smithville area, the central village in the subwatershed.

The Sugar Creek Watershed and the Upper Sugar Creek

The Sugar Creek watershed is located in northeast Ohio and covers 357 square miles across four different counties, including Wayne and Holmes, the leading dairy counties in the state. The mainstream is 45 miles long, stretching from north of the village of Smithville and empties into the Tuscarawas River near the city of Dover. The Sugar Creek lies within the headwaters of the Muskingum watershed, Ohio's largest hydrologic basin draining 8,051 square miles, or approximately one-fifth of the state. Although the watershed is relatively small in size, there is significant biophysical and social diversity across and within its six subwatersheds.

More than 70% of the land area in the watershed is in agricultural use, including the mix of row crops, dairy farms, and beef and poultry confined feeding operations that is typical of the Midwest and Mid-Atlantic regions. Yet, while the landscape within the Sugar Creek watershed can be characterized as largely agricultural, there are segments of second-growth forest, some riparian corridor, and several small-sized municipalities. In addition, farming practices in these six agricultural subwatersheds range from the traditional Amish farming practices found in the southern subwatersheds, to the conventional row crop production that is characteristic of the northern subwatersheds and is pervasive in the Upper Sugar Creek.

A Highly Degraded Agricultural Watershed

In 1998, the Ohio Environmental Protection Agency (EPA) conducted an assessment of the Sugar Creek watershed, concluding that the watershed is one of the two most degraded watersheds in the state. Four problem areas were identified: (1) excessive deposits of sediment in the streambed; (2) extensive stream channelization and riparian habitat destruction; (3) excessive nitrogen and phosphorus loadings; and (4) high levels of fecal coliform bacteria from both human and animal sources. Furthermore, carbon-cycling processes were negatively impacted due in large part to loss of riparian habitat and severe alteration of stream morphology.[2] The report identified agriculture as the principal source of watershed degradation.

Given the visibility of the problem of hypoxia in the Gulf of Mexico, concern with other environmental impacts throughout the Mississippi basin, and Ohio EPA's assessments of the Sugar Creek, it was no surprise that it was one of the first to undergo total maximum daily load (TMDL) planning in Ohio. The TMDL report summarized the 1998 survey in relation to the visible impact of agriculture on water quality as follows: "Observed aquatic resource degradation from agriculture included direct manure and urine discharge to streams, milking waste discharged by pipe to streams, dumping of fruit processing waste into streams, direct habitat alteration by dredging and cattle walking in streams, and lack of wooded riparian corridor."[3]

The Upper Sugar Creek subwatershed, which is the focus of this chapter and the headwaters of the Sugar Creek, is in Wayne County, Ohio, with the Village of Smithville at its center. More than 80% of the land in this part of the basin is devoted to agricultural activities; primarily grain farming that utilizes a 2-year corn–soybean rotation, followed by dairy and other forms of animal husbandry. The average total farm size, including leased land, is approximately 287 acres, and almost two-thirds of the farmers lease some land for production. Many of the farmers in the Sugar Creek, like farmers across the country, are responding to external market pressures by intensifying production and decreasing fallow cycles. In the Upper Sugar Creek, farmers are generally increasing their scale of operations, and, as a consequence, average acreages have almost doubled in the last 20 years.[4] In its assessment of the Upper

---

[2] Ohio Environmental Protection Agency. 2000. *Biological and Water Quality Study of Sugar Creek, 1998*. OEPA Technical Report MAS/1999-12-4.

[3] Ohio Environmental Protection Agency. 2002. *Total Maximum Daily Loads for the Sugar Creek Basin*. (http://www.epa.state.oh.us/dsw/tmdl/SugarCreek.html).

[4] Parker, Jason, Richard Moore, and Mark Weaver. 2007. "Land Tenure as a Variable in Community Based Watershed Projects: Some Lessons from the Sugar Creek Watershed, Wayne and Holmes Counties, Ohio." *Society and Natural Resources* 20(9):815–833.

Sugar Creek, the *Sugar Creek Watershed TMDL* concluded that this subwatershed suffers from essentially the same problems as the watershed as a whole, with agricultural activities the major sources of impairment.

Organization and Process: What Kind of Group is Appropriate?

A group of social scientists, natural scientists, and local collaborators affiliated with the Agroecosystems Management Program (AMP) at the Ohio Agricultural Research and Development Center (OARDC) began to address questions about the formation of a watershed group in 1999. AMP is an interdisciplinary program at the Ohio State University that focuses on the concept of agroecosystems, defined as the integration of production, environment, economics, and social systems. Participants include shareholders, students, staff, and teaching, research, and extension faculty from the natural, physical, economic, and social sciences at OSU and other academic institutions. AMP's shareholders include representatives of farmer, commodity, and environmental groups and governmental agencies. The Program's purpose is to improve the economic, environmental, and social viability of Ohio's agriculture and rural communities through ecological approaches. OARDC, OSU Extension, the College of Food, Agricultural and Environmental Sciences, and the W.K. Kellogg Foundation all provide support, along with many individuals and organizations. Figure 16.1 illustrates the multiple funding sources and organizations involved in participatory headwaters research associated with the Sugar Creek project.

Richard Moore, an anthropologist on the AMP team, drawing on his experience in studying Japanese agriculture and conservation practices, sought to initiate a project focused on the headwaters of a local watershed.[5] In addition, the AMP team made a decision to commit to a subwatershed approach as the only way to develop genuine grass roots participation. AMP was already working with local stakeholders, including a dairy farmer living in the Upper Sugar Creek subwatershed.

With agriculture identified as a principal source of water degradation, the AMP watershed team chose to focus on farmers, who owned most of the land along the stream. In addition, they decided to start an informal group of only farmers rather than a formal watershed group representing different stakeholder interests. The Sugar Creek watershed was experiencing the typical pattern of rapid exurban growth with new housing subdivisions occupied by residents who commuted to one of the several nearby cities. These new housing developments were dispersed in small clusters throughout the watershed, with increasing conflicts regarding farm

---

[5]For more detail on the formation of the Sugar Creek Partners and "the Sugar Creek method," see Moore, Richard, Jason Parker, and Mark Weaver. 2008. "Agricultural Sustainability, Water Pollution, and Governmental Regulations: Lessons from the Sugar Creek Farmers in Ohio." *Culture and Agriculture* 30(1–2):3–16 and Morton, Lois Wright and Steve Padgitt. 2004. "Selecting Socio-Economic Metrics for Watershed Management". *Environmental Monitoring and Assessment* 30:1–16.

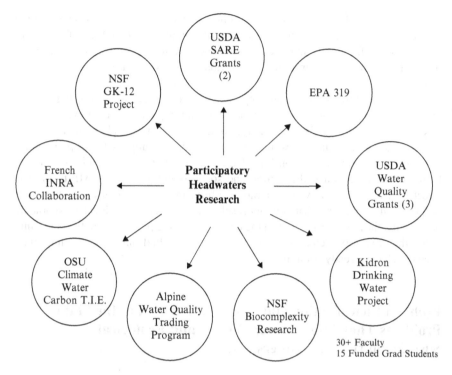

Fig. 16.1 Sugar Creek projects

machinery on the roads, noise, smells, and related issues, and increasing tension between farmers and nonfarmers in the local area.

The AMP team recruited a group of local producers who were neither the largest nor the most influential farmers in the area. Utilizing a grassroots "snowball method," the local farmer who was a member of AMP approached three other farmers who might be receptive to working together on water quality issues. Each of these farmers was asked to attend a local meeting and bring at least one additional neighboring farmer to determine if there was sufficient interest in forming a group and working with the AMP team.

The Sugar Creek Partners, which first met in September of 2000, meets monthly during the winter months, relying primarily on phone, email, and informal face-to-face contacts during the busier parts of the year. The meetings are informal, there is no committee structure, and there is no formal voting procedure. The Partners functions very much like a traditional farmers' "learning circle" in which participants share local knowledge and discuss information presented by scientific "experts" or representatives of local agencies.[6] The meetings are chaired by a

---

[6]Cranford, Elaine E. and Julie Kleinschmit. 2007. *Fishbowls in the Field: Using Listening to Join Farmers, Ranchers, and Educations in Advancing Sustainable Agriculture.* CARI: Center for Applied Rural Innovation at the University of Nebraska, Lincoln (http://digitalcommons.unl.edu/caripubs/39).

farmer who typically asks: "So are we all in agreement with going ahead and doing X?" to conclude the discussion of a particular item. The members of the AMP advisory team frequently make comments and recommendations, but the farmers make all the decisions through this informal, consensus-based decision-making process.

We found that this farmers-only approach was very effective in fostering a frank, open discussion of issues among the members and in providing them with a forum in which they were comfortable enough to identify, discuss, and pursue the watershed issues that they thought were most important. This approach was particularly effective in building a relationship of trust among the group members and also between the farmers and the natural and social scientists in the AMP team. The effectiveness of voluntary collaboration efforts in resource conservation has its skeptics, but we found that this approach can be effective if local stakeholders believe that they are participants in the decision-making process from the beginning rather than token participants whose views are consulted only after the important decisions have already been made.

Problem Identification and Definition: What are the Main Problems That Affect Water Quality and Ecological Sustainability in the Watershed?

One main obstacle lay in the way of recruiting and organizing grassroots stakeholders in the Upper Sugar Creek: many local residents, including farmers, were not aware of or did not recognize the water quality issues as "problems." Although Upper Sugar Creek had been assessed as impaired, from the beginning, the farmers made it clear that they did not trust Ohio EPA or what Ohio EPA scientists said about "their stream." This reaction was expected because the AMP team had conducted a social survey of all property owners whose land abutted the Upper Sugar Creek, and found that only 11% of all respondents (including nonfarmers) agreed or strongly agreed with the statement that the Sugar Creek is polluted.[7] In contrast, 46% disagreed or strongly disagreed, and 41% indicated that they neither agreed nor disagreed. In other words, in this subwatershed there was clearly a major disconnect between the assessment of the water quality by Ohio EPA scientists and the perception of the landowners living along the stream. This problem made group formation more difficult than is the case in watersheds where groups have already organized around conflicts over the resource or where the local residents are well aware of existing problems.

---

[7] For more information on the survey methodology and results, see Weaver, Mark, Richard Moore, and Jason Parker. 2005. Understanding Grassroots Stakeholders and Grassroots Stakeholder Groups: The View from the Grassroots in the Upper Sugar Creek, presented at the 2005 Annual Meeting of the American Political Science Association, Washington, September 1–4 (http://convention2.allacademic.com/one/apsa/apsa05/).

In addition, the AMP survey indicated that property owners living along the stream had little trust in EPA or Ohio EPA. Respondents were asked to rank on a 6-point scale, with 0 being "do not trust" and 5 being "trust very much," 13 agencies and organizations. US EPA and Ohio EPA had the lowest mean scores, 1.67 and 1.86, respectively. The landowners in the Upper Sugar Creek expressed higher trust in local, nonregulatory agencies as opposed to federal or state agencies or agencies with taxing or regulatory power. But the lack of trust in Ohio EPA carried over to what Ohio EPA scientists reported regarding impairments in the Sugar Creek, and this lack of trust posed additional difficulties in getting the farmers to accept that there were problems that needed to be addressed.

Furthermore, the farmers did not simply voice a general lack of trust in Ohio EPA, but rather questioned and challenged the 1998 data, the methodology used to collect the data, and Ohio EPA's conclusion that agricultural practices were the primary contributor to water quality impairments. Through the first several meetings of the Partners, the farmers voiced several challenges to EPA's study. They were never consulted during the collection of the data. They criticized the lack of flow data and questioned whether the data had been collected shortly after major rain events. They stated their belief that the sampling sites were too few and the collection process not long enough (Ohio EPA sampled four sites in the Upper Sugar Creek, once per month for 5 months). They challenged Ohio EPA's logic in concluding that agriculture was primarily responsible for stream impairment based upon the evidence presented. And they questioned Ohio EPA's assessment of the extent of impairment, especially the characterization of the Sugar Creek as the second most degraded watershed in Ohio (second only to the Cuyahoga, which is infamous for catching on fire).

The AMP team engaged the Partners in a serious discussion of the criticisms raised by the farmers. While agreeing that the EPA findings were based on sound science, the AMP team still focused most of the early meetings on trying to translate Ohio EPA's data and conclusions into terms that the farmers could understand, and serving as a mediator to take farmers' questions to EPA scientists and to take the answers back to the Partners. However, although the Partners achieved a better understanding of the EPA findings, they remained distrustful of data that they could not independently verify.

When offered the option of working with the AMP team to conduct an independent analysis of water quality, the Partners decided that this was the only way that they could determine whether the Upper Sugar Creek was really as polluted as Ohio EPA claimed. As a result, members of the AMP team began biweekly sampling of 21 sites in the nonwinter months. Personnel from Wayne County Soil and Water, AMP, and the Sugar Creek Project, once it became independent of AMP, conducted the ongoing water sampling, which has been funded by AMP, a 319 Grant from EPA, and a variety of other funding sources. The farmers helped to determine the location of the sampling sites, with a focus on the land near their farms and homes. This decision was based on their desire to sample as close to their own farms as possible, and to avoid being associated with a sampling program that suggested they were interfering in the affairs of landowners outside the group. From the beginning of the sampling, it was clear that the data collected by the local watershed group largely confirmed Ohio EPA's assessment of the impaired condition of the Sugar Creek.

Causation and Responsibility: What Are the Causes of These Problems, and What Individuals or Groups Bear Primary Responsibility for the Impairments That Exist?

Although the water sampling data collected with their participation convinced the Partners that there was a problem, determining ultimate causation and responsibility remained impossible. The sources of excessive siltation clearly included erosion from numerous home construction sites and other nonfarm sources in the subwatershed as well as agricultural run-off. Similarly, both the nutrient loading and the high bacteria counts had nonfarm as well as farm sources.

In their discussion of the proposal to do independent sampling, the farmers acknowledged their responsibility to ensure that their operations were not polluting the stream, and they indicated that they would take steps to remediate those causes of impairment that could be traced to their farms. However, it soon became clear that there would be no definitive, scientifically based determination of the causes of the problem upon which some type of fair division of remediation efforts between farm and nonfarm sources could be established. Such an inability to identify causes and assign responsibility is frequently a major obstacle to organizing and implementing conservation efforts. However, despite uncertainly about the causes of the problem and who was primarily responsible, the Partners decided to take action by implementing conservation practices. In other words, although debates over the causes of problems and the assignment of blame or responsibility are a dominant feature of US public policy in general and resource policy in particular,[8] the Partners chose to take responsibility rather than to play the blame game. This choice was a critical point in the decision-making process of the Partners, because it occurred at precisely the moment when social scientists would expect a group to use the uncertainty over causation and inability to assign responsibility to justify a course of inaction.

In part, this decision reflected a largely unarticulated set of fundamental beliefs about stewardship of the land and water and a sense of responsibility to neighbors and the local community. From the beginning, the farmers had indicated that it was their responsibility to ensure that the water in the Sugar Creek should be as clean when it left their land as when it entered their land. Several members of the Partners were extremely embarrassed when the local newspaper carried the headline that the Sugar Creek was one of the most polluted streams in the state. They exhibited great

---

[8]Stone, Deborah. 2001. *Policy Paradox: The Art of Political Decision Making*. New York: W.W. Norton & Co.

pride in their role as stewards of the land, and they clearly wanted the rest of the community to see them as good stewards of the land. This powerful sense of stewardship was a major factor in the Partner's decision to participate in conservation and remediation activities despite uncertainty over the causes of the water quality problems in the Upper Sugar Creek.

Participation: Whose Participation Will Be Sought in Working to Solve the Problems?

The Sugar Creek Partners adopted an informal and low-key approach to inviting additional participation in conservation and remediation efforts. For example, in July of 2001, the 12 farmers in the Partners agreed to send letters to all adjacent landowners who owned more than 10 acres to inform them of the Partner's activities. These letters had the purpose of informing neighbors about the ongoing water sampling program and other Partner activities, and of indirectly encouraging additional participation in conservation practices, whether as individuals or as part of the group. This approach reflected the Partner's vision of themselves as a catalyst of change rather than any kind of attempt to take charge of overall planning and management in the watershed.

The farmers discussed the possibility of recruiting members from outside the farming community and decided to remain an all-farmer group. This decision did not reflect a desire to retain some kind of control over watershed planning or remediation activity, since the group had no official or even semiofficial status as a policy making group or as an advisory group to policy makers. The Partners believed that they could be most effective as a farm group that would encourage volunteer efforts to positively impact the watershed. They decided to reach out to the community and inform others about water quality issues and to encourage participation in conservation practices through a variety of public education efforts that will be described in the next section.

The Partners did initiate one type of direct approach to recruiting additional participants in the remediation efforts as a result of the sampling program. Although the water quality testing could not identify the causes of nutrient loading, it could identify particular segments of the Upper Sugar Creek as having especially high nutrient loadings. One of these problem areas or "hot spots" was located in a small stream segment where there was very high probability that the excessive nutrients originated either in one dairy farm or from a small set of nearby houses. The Partners agreed that members of the AMP team should initiate contact with the farmer and house owners and present them with the sampling data.

In the initial contact by AMP team members, the house owners expressed the view that the excess nutrients must be from farm run-off, while the farmer was just as certain that the problem was with aging septic systems in the houses. The homeowners agreed to allow dye tests of their septic systems, which came out negative. After the dye testing eliminated the houses as sources, two AMP researchers with the backing of the Partners again approached the dairy farm owners. With the researchers' support, the dairy farmers agreed to document the extent of the problem and to develop a solution. A farm-wide nutrient budget was developed for improved manure management, application of manure was redirected, and cattle access to the stream was restricted. The Partners saw this outcome as one more demonstration of their approach, which encouraged the voluntary participation of other grassroots stakeholders, whether farmers or nonfarmers, as the best type of solution.

Actions and Solutions: What Types of Actions or Solutions Will Be Pursued By the Group?

Individual members of the Partners initiated numerous changes on their farms despite their uncertainty about the sources of pollution noted above. Members of the group created an 8-mile buffer strip to reduce pollution from agricultural runoff and protect against stream bank erosion. They also planted more than 1,200 trees in riparian zones and initiated several new "best management practices" including grass waterways and a wetland common across several properties. In addition, some of the Partners adopted more ecologically oriented management systems, including such practices as conservation tillage, crop rotations, contour strip cropping, cover and green manure crops, filter strips, grasses and legumes in rotation, fencing out cows, improved pasture, riparian forest buffer restoration, and manure management.

The Partners consciously attempted to include the families of the participants in learning about the stream, impairments, and remediation efforts. For several years, the Partners held a family day that combined a picnic outing at a location along the stream with some type of demonstration or presentation of information on the watershed. Family day always included some activities designed to engage the children in stream conservation, and some activity that "gets people into the stream." In addition, although they seldom attended the monthly meetings, the wives of the Partners were actively involved in the decision making regarding conservation changes on each farm. In short, the Partners were successful in involving the families of the participants in their activities.

The Partners also became involved in public education outreach aimed at other farmers, schoolchildren in the watershed, and the wider community. Each year the group sponsored a farm tour to demonstrate some of the changes that individual members had instituted. Other members of the Partners became involved in watershed educational programs that were introduced into the local schools and focused on the local watershed. The AMP team and the Partners also worked together on projects to inform the wider community, such as sending a newsletter focusing on issues relating to the watershed to all residents in the Upper Sugar Creek; building a kiosk in the park of the largest village in the subwatershed to provide information on the watershed and the Partner's activities; and providing well water testing to increase awareness of water quality issues.

In a 5-year period, the Partners grew to a group of 17 participating farmers who all have land along the stream and who collectively manage approximately 40% of the agricultural land in the Upper Sugar Creek basin.

Values and Goals: What Is the Basic Vision or Set of Value Commitments That Will Guide the Group?

The AMP team did not attempt to lead the farmers in the group in any kind of "visioning" exercise and did not encourage the Partners to formulate a formal subwatershed plan with explicit goals and value commitments. From the beginning, the Partners meetings had a very pragmatic, problem-oriented approach, without any real attempt to articulate a common vision or mission statement for the group. In part, this reluctance to formulate an "official" watershed vision or mission statement reflected the Partners' strong desire to avoid being seen as making decisions for or speaking for others. They never portrayed themselves as the representatives of other farmers or of the community at large, even though they did see themselves as taking actions that benefited the farming community and the wider community. There was also an implicit but strong understanding that individual members would decide what conservation practices to implement on a voluntary basis, and that group decisions bound individuals only to the extent that each individual agreed to be bound by the group.

Of course, this focus on problem solving did not mean that the Partners' meetings never addressed issues involving values or vision. For example, one member argued against the notion of buffers as a solution, stating that the goal should be to have whole farm management practices that caused no problems in water quality rather than buffering farms away from the stream. Another Partner proposed the creation of a contiguous riparian corridor all along the mainstream that might be used for recreation or other community purposes. Yet, despite these discussions, the Partners never expressed the need or desire to attempt to articulate a mission statement or watershed vision that would guide the group as a whole or communicate their vision to the wider community.

As noted above, we believe that the Partners avoidance of defining their group in terms of a mission statement or watershed plan reflected a different kind of approach to watershed "management." Because they perceived themselves as catalysts of voluntary, local-based solutions, they avoided the standard language of watershed management, which for them remained tied to regulatory, top-down, centralized approaches to problems. Moreover, we believe that such a standard management approach to watershed management was itself anathema *to the common values and collective vision of the surrounding farm community.*

The member of the Partners who initiated the most conservation measures and was given a county conservation award, summarized the basis of his commitment to conservation measures in a 2004 newspaper interview in this way: "As a Christian, I feel that being a good steward of the land is very important. We have to protect our soil and our water."[9] However, none of the Partners, including this farmer, ever felt the need to discuss their Christian values in relation to what they should or should not do in the watershed or on their farms. The members of the Partners shared the common language and social bond of the culture and values of the local farm community, and they accepted the AMP team as working within this framework. They would have reacted very differently to, and probably refused to collaborate with, a group of scientists or organizers who spoke in terms of a management model that failed to engage *the core values of their community*.

The Gap in Watershed Management Language and Assumptions

To elucidate the gap between the language and assumptions of watershed "management" and the local farm culture in the Sugar Creek, we draw on a series of lengthy, open-ended interviews with farm families in the area.[10] In May through July of 2005, three interviewers completed 30 open-ended interviews of residents of the Smithville area. The interviews, which ranged from 45 min to longer than 2 h in length, were conducted as oral histories that allowed the interviewees to describe their lives in their own words and in terms of the categories and concerns they considered most important. Although the researchers invited participation from both husband and wife when making contact and even offered separate interviews, only five couples agreed to be interviewed, all in joint interviews. These interviews, which focused on the history of farm families in the area, including families inside and outside the Partners, provide a snapshot of the core values of this local culture. The farmers interviewed in the study clearly did see themselves as part of a locally based agricultural community or farming community. The exact boundaries between "we" and "they" were often unclear because of the family members who left farming, but the core beliefs and values that define the "we" of the social identities of these grassroots stakeholders were evident in the narratives they told. While there were many differences in their family narratives, four major themes were pervasive.

First, the farm families believed that they have a more independent lifestyle than the nonfarm community, and that an entire community of like-minded individuals and farm families supports this independent lifestyle. Moreover, they consistently maintained that "government" represented the greatest threat to this value of

---

[9]Downing, Bob. 2004. Farmers Help Sugar Creek Grow Cleaner. *Akron Beacon Journal* 24:B1.

[10]For more information on the methodology and summary of these interviews, see Weaver, Mark, Richard Moore, and Jason Parker. 2005. Understanding Grassroots Stakeholders and Grassroots Stakeholder Groups: The View from the Grassroots in the Upper Sugar Creek, presented at the 2005 Annual Meeting of the American Political Science Association, Washington, September 1–4 (http://convention2.allacademic.com/one/apsa/apsa05/).

independence that they consistently placed at the center of their lifestyle, making them automatically suspicious of "government" involvement in their lives.

Second, they expressed the shared belief that the farming lifestyle is unique in contemporary American materialist culture because it values hard, physical work and offers rewards, such as a life lived close to rather than apart from nature, beyond those of a merely material life. In addition, they valued this demanding farming lifestyle because it produces strong families and children with moral values, in contrast to what they saw as the dominant developments, including loss of contact with nature and obsession with material things in the culture at large.

Third, they perceived current patterns of "development," generally characterized in terms of the invasion of exurban sprawl into the area, as posing a threat to the continuation of the family farm and to the survival of the farming community. Many of the farm families specifically identified a farmer's decision to sell land as a moral as well as an economic decision that has an impact on the entire farming community.

Fourth, they linked farming to faith and saw the local farming community as grounded in Christian values despite recognized differences among the sects by which individuals and families were identified to the interviewers. Once again, they drew a clear line between "we" and "they" in describing these Christian values as increasingly ridiculed and rejected in secular institutions such as the public schools and by the growing secularism of the wider culture.

These core beliefs and values of the local farm culture are not simply an abstract set of philosophical commitments, but rather constitute part of their social identities of who they are. These core beliefs and values are strong psychological attachments that link individuals and families to the local farming community, and they are the common lenses through which they see and understand new issues that are brought to their attention. As the Partners came to understand the impairment of the Upper Sugar Creek, they saw their responsibilities and role in terms of these cultural values. The AMP team was able to collaborate with the Partners successfully because the farmers perceived them as sharing similar values and beliefs rather than as representatives of a hostile outside culture.

Learning from the Sugar Creek Partners Process

Organization and Process

Voluntary collaboration among farmers and other resource users is not the silver bullet of watershed management; however, it can be effective under certain circumstances. The organization and process of the proposed watershed group should not simply follow a textbook model, but rather be designed such that it is user-friendly to those grassroots stakeholders whose collaboration is required.

Problem Identification and Definition

Problem identification and problem definition pose difficulties in watersheds where the problems of impairment are only visible downstream in the eyes of most stakeholders. In such cases, local stakeholders are likely to distrust outside experts, including scientific experts and public officials, and they tend to distrust data on "their stream" that they cannot independently verify.

Causation and Responsibility

Establishing causation and responsibility is typically a critical step in generating collective action to address an environmental problem. This step is notoriously difficult when addressing nonpoint source pollution problems. Because of this, watershed organizers should foster a stakeholder discussion that develops their own sense of responsibility and stewardship in the watershed. Moreover, an approach that includes assignment of blame generates animosity and erects social barriers that can lead to noncooperation of those community members whose cooperation is needed to make the project a success.

Participation

A watershed group that does not make public policy does not need to be based on the pluralist model that requires the participation of every stakeholder interest in the watershed. Different modes of engaging stakeholder participation can be effective dependent upon local circumstances.

Actions and Solutions

The goal of comprehensive watershed planning that coordinates conservation and remediation efforts should avoid drowning out the benefits of local, smaller scale actions and solutions. Where the participation of grassroots stakeholders is essential, the comprehensive plan needs to be built from the bottom up rather than imposed from the top down.

Values and Goals

Local agricultural stakeholders typically believe that they do not share values with and are misunderstood and misrepresented by environmentalists, governmental officials, and the public at large. Gaining local collaborative support of landowners and natural resource users is often critical to effective watershed management. Organizers of grassroots watershed groups must be willing and able to speak about their purposes and goals in terms that address the core beliefs and values of the local community.

Acknowledgment The authors express their gratitude to the farmers and residents of the Sugar Creek watershed, especially the Sugar Creek Partners, the North Fork Task Force, and members of the Amish Church Districts. We also thank the Holmes and Wayne County Soil and Water Conservation Districts, Natural Resources Conservation Service, Wayne County Auditor, and Wayne County Extension for their support and participation in this research. This research was funded by grants from the United States Department of Agriculture Sustainable Agriculture Research and Education program and the United States EPA.

Chapter 17
Farmer Decision Makers: What Are They Thinking?

Lois Wright Morton

...reasons we can't get away from fall tillage. Farmers have ... reasons ... social, because you've got the neighbors out tilling; two I don't know what's going to happen next spring, but I've got good weather now ... and three I can pay $570 a ton for my anhydrous vs. $690 a ton come spring.

Iowa Learning Farm farmer 2008 interview 02120805

My neighbor tells me that he thinks that unless he moldboards he doesn't get a good mix in his topsoil of getting his fertilizer and residue mixed in and do what it's supposed to do.

Iowa Learning Farm farmer 2008 interview 02080802

...my grandfather prided himself on very straight rows ... and clean plowing ... when I'm growing up, if you had any cornstalks showing, you were doing a poor job. I mean you were judged on how good a job you could do getting all that stuff rolled over and making it perfectly black.

Iowa Learning Farm farmer 2008 interview 02010817

The perception [is] that it just doesn't look as good up there when that residue is standing up.

Iowa Learning Farm farmer 2008 02010810

Conservation Agriculture

A farmer's decision to practice conservation agriculture is not just a decision to adopt one new farming technique but a reconstruction of beliefs and values about the agroecosystem and their role as manager of a new crop production system.[1] Concerns over water quality, air quality, and other environmental issues

---

[1]Coughenour, C. Milton. 2003. "Innovating Conservation Agriculture: The Case of No-Till Cropping" *Rural Sociology* 68(2):278–304.

L.W. Morton (✉)
Department of Sociology, Iowa State University, 317C East Hall, Ames IA 50011-1070, USA
e-mail: lwmorton@iastate.edu

L.W. Morton and S.S. Brown (eds.), *Pathways for Getting to Better Water Quality: The Citizen Effect*, DOI 10.1007/978-1-4419-7282-8_17,
© Springer Science+Business Media, LLC 2011

have radically expanded the mandate of conservation agriculture in the past 20 years compared with conventional soil conservation practices in most of the twenty-first century. Conservation agriculture today denotes a new paradigm requiring new value orientations, new identities, and mental maps about the environment and the production of agricultural products.

If the agricultural landscape across the United States is expected to co-produce agricultural products and ecosystem services, more farmers must make the decision to become conservation farmers in this new, expanded sense of environmental managers. Scientists, educators, and agency technical specialists have treated the problem of conservation technology adoption as a knowledge transfer from scientist to farmer. Underlying this knowledge transfer process is the assumption that the scientist is providing a superior substitute for the customary managerial expertise of the farmer.[2] This expert conception of how farmers ought to make decisions is a barrier to farmers' learning new systems and adopting conservation measures that are meaningful and relevant to them.

McCown's[2] essay on "New Thinking About Farmer Decision Makers" is used to frame the discussion in this chapter about how farmers make decisions under conditions of ambiguity and uncertainty. Expert and customary management and the assumptions that influence differences in how the scientist and educator vs. the farmer think about management are compared. Then the farmers' perspective and how experiences, knowledge, and social relationships create mental maps used to make sense of the world around them are presented.[3] Data from farmer surveys are used to illustrate these concepts and what farmer decision models might look like. The chapter ends by challenging the reader not to make assumptions about what farmers are thinking, but to take time to ask them and listen to the answers.

Data sources referenced in this chapter come from multiple sources, including 2006 key informant surveys of 360 conservation-minded farmers and 358 extension and natural resource professionals in 68 randomly selected Hydrological Unit Code (HUC) 12 watersheds in Iowa. Additional data are from a 2007 survey of the Lower Big Sioux River in northwest Iowa ($N = 1,110$ farmers and community members; 11 subwatersheds), six in-person audiotaped interviews with northeast Iowa farmers in 2005, and three 2008 audiotaped listening sessions conducted by the Iowa Learning Farm with farmer cooperators.

---

[2]McCown, R. L. 2005. "New Thinking About Farmer Decision Makers." In *The Farmer's Decision: Balancing Economic Successful Agriculture Production with Environmental Quality* edited by Jerry L. Hatfield. Ankeny: Soil and Water Conservation Society.

[3]Ryan, Robert L., Donna L. Erickson, and Raymond deYoung. 2003. "Farmers' Motivations for Adopting Conservation Practices along Riparian Zones in a Midwestern Agricultural Watershed." *Journal of Environmental Planning and Management* 46(1):19–37; Napier, T. L., M. Tucker, and S. McCarter. 2000. "Adoption of Conservation Production Systems in Three Midwest Watersheds." *Journal of Soil and Water Conservation* 55(2):123–134.

Facilitative Learning Rather than Prescriptive

The mechanization and modernization of agriculture based on the redesign by experts on how labor activities could be organized for increased efficiency followed closely their application to US manufacturing industry.[2] This new modern paradigm replaced the manager as craftsman with the manager as technician. "Scientific principles for management came to be viewed as superior to customary managerial expertise" and interventions in influencing management behavior presupposed expert knowledge as a superior substitute for problem solving.[2] Further, expert scientists and agronomic educators considered the intuitive judgment of the farm manager as a problem to be overcome and when possible replaced with their recommended systems of formal rational analysis and decision making. Classic economics reinforces the idea of the rational manager who has perfect knowledge and a goal of profit maximization that can be easily achieved by making rational choices.[4]

Under these assumptions, the information transfer strategy of the scientific and agricultural education community becomes one of investing in technology. The expectation is that the rational farmer will welcome science and information products that are relevant to routine practice and easily accessible. While many agricultural innovations, such as hybrid seeds, genetically modified plants, and GPS-guided equipment, have been widely adopted, conservation management practices have lagged considerably. McCown[2] laments the low rate of farmer adoption of innovations in management such as agricultural decision support systems for insect control that uses a logic system proven to be sound but never adopted by more than 30% of Australian cotton growers. He further notes that, in the end, routine use of the decision support software was replaced by farmers' monitoring and subjective judgment with the decision support system *facilitating* learning and guiding rather than prescribing management decisions.

It seems that indeed conservation agriculture is qualitatively different than conventional agriculture.[1] The innovation-diffusion model by Rogers[5] is often given as explanation for adoption of new practices in agriculture. However, a number of social scientists have challenged this model, finding it is "not even close" to predicting farmers' conservation behaviors.[6] One reason may be that conservation management is not

---

[4]Simon, Herbert. 1979. "From the Substantive to Procedural Rationality." pp. 65–68 in *Philosophy and Economy Theory* edited by F. Han and M. Hollis. New York: Oxford University Press.

[5]The innovation-decision model consists of five stages: knowledge, persuasion, decision, implementation, and confirmation. Rogers, Everett M. 1962, 1971, 1983, 1995, 2003. *Diffusion of Innovations*. New York: Free Press.

[6]Nowak, Pete and Peter F. Korsching. 1998. "The Human Dimension of Soil and Water Conservation: A Historical and Methodological Perspective." In *Advances in Soil and Water Conservation* edited by F. J. Pierce and W. W. Frye. Chelsea: Ann Arbor Press; Lockeretz, William. 1990. "What Have We Learned About Who Conserves Soil?" *Journal of Soil and Water Conservation* 445(1):132–136; Parker, Jason Shaw and Richard H. Moore. 2008 "Conservation Use and Quality of Life in a Rural Community: An Extension of Goldschmidt's Findings." *Southern Rural Sociology* 23(1)235–265.

a single practice or technology but a system of management that requires judgment to match appropriate practices to conditions. The conservation farmer does not have a technical script regarding what should be done but rather must develop a technical frame of mind, a set of knowledge, methods, and routines. Many environmental management techniques are complex, without standard procedures, and thus need managers willing to experiment, take risks, and learn what works best under each unique set of conditions of soil, topography, crop, weather, and other variables.[7]

Managing the agricultural landscape is as much art as science and decisions cannot be reduced to explicit prescriptive rules to guide all situations. New ecological theory describes nature as less stable than previously modeled, nonlinear, and full of surprise.[8] If this is true, these dynamic natural systems require management that is flexible and managers who combine customary knowledge built from experiences and applied innovative technologies evaluated through an experimentation, monitoring, learning, and adaptation cycle.

Those who embrace conservation management must shift their thinking from resource extraction goals in the production of agricultural products to some balance between production and sustaining the health and productivity of their natural systems. It is not that ecological integrity trumps other human goals, but it must be realized that there is no way to sustain human society without sustaining soil and water resources.[8]

The Farmer's Perspective

Learning takes place when people discover for themselves contradictions between observed behavior and their perceptions of how the "world" should operate.[9]

Farmers' mental maps about their farm, their watershed, environmental and water quality issues, and agricultural and environmental goals are often different than those of scientists and technical professionals.[10] The term mental map means the conceptual frame a person carries in their mind to explain the way the world

---

[7] van Es, J.C and P. Notier. 1988. "No-till Farming in the United States: Research and Policy Environment in the Development and Utilization of an Innovation." *Society and Natural Resources* 1:93–110.

[8] Grumbine, R. Edward. 1997. "Reflections on 'What is Ecosystem Management?'" *Conservation Biology* 11(1):41–47.

[9] Morecroft, John D. W. 1994. "Executive Knowledge, Models, and Learning." p. 4, Chapter 1 in *Modeling for Learning Organizations* edited by John D. W. Morecroft and John D. Sterman. Portland: Productivity Press.

[10] Christensen, Lee A. and Patricia E. Norris. 1983. "Soil Conservation and Water Quality Improvement: What Farmers Think." *Journal of Soil and Water Conservation* 38:15–20; Morton, Lois Wright. 2008. "The Role of Civic Structure in Achieving Performance-based Watershed Management." *Society and Natural Resources* 21(9):751–766.

works.[9] This map is a network of information, facts, subjective experiences, ideas, and observations from which opinions, options, courses of action, and expectations about the outcomes of management decisions are derived.[9] The quality and accuracy of these maps (e.g., how well they mimic reality) affect the farmer's capacity to manage for profitability and water quality goals.

Traditionally the farmer's role in society is to produce food, feed, fuel, and fiber for human uses. The ecological scientist and the watershed specialist roles are focused on protecting and restoring the ecosystem and natural resource base. The nonscientist often evaluates interactions with the environment from their life experience and relevance to personal situations that, in turn, generate local knowledge.[11] The scientist relies on the process of scientific inquiry and measured data to create a different kind of knowledge. Each has access to different kinds of information and filters information based on their roles; assigning different interpretations and meanings to available scientific facts. These worldview differences are evident in surveys that compare farmers, scientists, and agency natural resource professionals.[8]

Scientists, watershed specialists, and environmental leaders are often frustrated that farmers do not respond to what they see as a crisis in water issues. Differences in perceptions are illustrated by a comparison of viewpoints from a 2006 Iowa survey of conservation-minded farmers (identified by extension and agency personnel) and natural resource professionals and extension educators (Table 17.1). The extension and the natural resource professionals in this sample were significantly more likely to believe that a wide variety of contaminants, such as nitrates, phosphorous, insecticides, and fecal coliform bacteria, were more serious problems than the

Table 17.1 Do you think the following threatens groundwater/ surface water quality in Iowa?

	Farmers	Agency/Comm.
	Mean	Mean
Nitrates	2.4	2.9***
Phosphorous	2.2	2.8***
Inorganic contaminants	1.9	2.1***
Insecticides	2.3	2.7***
Herbicides	2.3	2.7***
Soil erosion	2.6	3.1***
Fecal coliform	2.3	2.8***
N in your watershed	2.1	2.6***
P in your watershed	1.9	2.5***

Iowa conservation farmers ($N=360$); Natural resource/extension professionals ($N=358$) in 68 HUC 12 watersheds 2006.
1 = no threat/no problem; 2 = some threat/slight problem; 3 = serious threat/serious problem; 4 = very serious threat/problem.
***$p=<0.001$ significance

[11]Rhoads, Bruce L., David Wilson, Michael Urban, and Edwin E. Herricks. 1999. "Interaction Between Scientists and Nonscientists in Community-Based Watershed Management: Emergence of the Concept of Stream Naturalization." *Environmental Management* 24(3)297–308.

conservation farmer respondents. Further, farmer assessments of nitrogen (N) and phosphorous (P) as threats to their own watershed were much lower than those of natural resource "experts."

This finding suggests that the mental maps of these two groups are not the same. They likely have access to different kinds of information and are processing the meaning and relevance differently. These disjunctures, although seemingly small when comparing Table 17.1 means, can be the difference between awareness (some threat/slight problem) and motivation (serious threat/serious problem) to take action. Farmers in this sample may acknowledge water quality issues as potentially problematic but mentally are processing the information as not needing their attention or action vs. immediately problematic and needing their full attention.

This is a critical point. Until farmers (and people in general) believe there is a problem and that they should do something different, they will not be motivated to change their behaviors. They will not encourage others to do something different. It is not enough that expert opinion has defined the problem and it is even more unacceptable that experts have prescribed the solution for an unrecognized problem. McCown[2] and others assert that it is only when a situation is perceived as problematic that deliberation and purposeful thinking occurs and intervention becomes relevant. From a managerial standpoint, this response is appropriate. In normal, routine activities, decisions and actions are made based on past practices and not new conscious deliberation. This automatic behavior is an efficiency strategy, reducing the time and resources needed to manage day-to-day activities.

McCown[12] observes from nine case studies of decision support systems (DSS) applications that family farmers resisted replacing their decision processes with elaborate expert decision systems. When DSS were offered as tools to help farmers make decisions, they had more acceptance. Acceptance was not about the program but the manner in which it was given to the farmer. When offered as a prescriptive expert driven technology, it was less accepted than when agricultural educators engaged farmers in the problem and suggested DSS as one of many tools for solving the problem. The focus was on facilitating farmer learning rather than his or her adoption of a particular tool.

A typical educator/technical professional concept of intervention is, "I have the science and the technology, you just have to do what I tell you." This expert prescriptive intervention ignores and subtly disrespects farmers' preferences, experiences, and judgment as integral to their decision making. Surveys of farmers reveal that they trust their own judgment far more than that of an expert. Table 17.2 illustrates this point. When asked what the major factor was in making a manure application decision, the dominant response was "use own judgment based on experience." This experience may have included soil testing and assessment of crop nutrient requirements, but these pieces of information merely facilitate the decision rather than determine it.

---

[12]McCown, R. L. 2002. "Changing Systems of Supporting Farmers' Decisions: Problems, Paradigms, and Prospects." *Agricultural Systems* 74:179–220.

Table 17.2 Farmers give priority to their own judgment when they make decisions

When you apply manure, what is the *major* factor you use to determine application rates? *(Please circle only one.)* (n=225)

Crop nutrient requirements	29.3%
Ease of application	4.0%
Use own judgment based on experience	37.8%
Use manure sample	4.0%
Use soil tests	11.6%
Follow spreader manufacturer's recommendations	1.3%
Follow recommendations from agricultural scientists	1.3%
Pay little or no attention to application rate	4.4%
Follow consultant's recommendation	6.2%

Lower Big Sioux River NW Iowa (N=1,110 farmers and community members; 11 subwatersheds)

What Does the Farmer Decision-Making Model Look Like?

How does the farmer produce conservation behaviors? How can the educator and technical specialist help farmers reconstruct their mental maps of production agriculture to include conservation and environmental protection? McCown[2] advises that interventions must build on an understanding of farmers' customary management practices. Farmers are daily solving the problems of managing their agricultural enterprise. They must respond to complex interactions among production and environmental conditions, including the routine and the unexpected: changes in the weather, the seasons, commodity prices, disease and insect pressures, availability and skill level of labor, equipment condition, field soils and slope characteristics, personal time and energy, and impacts of previous decisions, to name only a few variables. When operating in such a complex system, no single change is without significant additional management and external costs. Farmers, like all humans, develop over time a set of automatic behaviors to accomplish and integrate the many tasks that must happen on a day-to-day basis. These routine decisions, when first instituted, likely were conscious decisions. The manager will implement these routine decisions as long as it seems that conditions have not changed and the management practice is still applicable. It is too costly to do otherwise.

Management decisions are grounded in shared knowledge of a community of practice (often passed from father to son and neighbor to neighbor) and individual expertise developed through experience. Everyday knowledge and actions are regulated by pragmatic motives and criteria such as "good enough," "just in time," and routinization of habits and recipes that work.[2] The role of planning and systematic decision making is illustrated in Table 17.3.

Manure management is a daily routine in animal agriculture. The sheer magnitude of manure accumulation requires a systematic plan so the farmer is not constantly taking time to gather new data, consider new options, and make a new decision. Northwest Iowa farmers' responses reflect the customary management practices and

Table 17.3 Farmers asked about their manure management (customary management)

How do you decide where to apply manure? *(Check all that apply.)*	
According to my manure management plan ($n=259$)	44.0%
Systematically rotate applications depending upon soil nutrient needs ($n=261$)	70.1%
Apply mostly in fields near my livestock facilities ($n=262$)	26.7%
Apply manure evenly in most or all of my fields ($n=260$)	50.8%
Apply in most convenient locations ($n=261$)	14.6%
Apply according to schedule that involves rotation of fields ($n=262$)	72.1%
Consultant's recommendation ($n=260$)	25.4%

Lower Big Sioux River NW Iowa ($N=1,110$ farmers and community members; 11 subwatersheds)

systems in place, with most rotating manure applications among fields depending on soil nutrients and according to a planned schedule. Of interest to the agronomic specialist concerned about excessive nutrients leaking into nearby water bodies, is that almost 51% of this sample apply manure evenly in most or all fields and a little more than a quarter apply mostly in fields near livestock facilities. The management assumptions behind this behavior would be that all fields have the same nutrient needs and/or that labor costs need to be minimized. These farm managers are missing the implications of what too much P or N on a field does to their water quality and production input costs.

Automatic Behaviors

Automatic management behaviors are efficient and practical. None of us reevaluate every decision we make every day. Most of the time we use "reference schema" to guide management decisions and behaviors. The mind uses past experiences and decision frames as rules of thumb to guide new decisions. The cognitive paradigm looks like this, "If there is a problem, try to arrive at a solution by applying a familiar (and simple) method, even if it does not seem entirely appropriate. When difficulties arise, try to overcome them by adapting the procedure in such a way as to make the difficulties disappear."[13] DeMay[13] continues,

> ...typically, problem solving within the frame approach avoids elaborate exploration of and evaluation of alternative methods for the solution of a particular problem. Again, the risk of drowning in complexity when trying to take into account every possibility and eventuality is estimated to be very high. Therefore a strategy of "ruthless generalization" rather than careful investigation is followed. On the basis of vague similarities with a familiar problem, a new problem is tackled with the procedure associated with the familiar problem, relying on debugging knowledge for making the procedure fit the new case.

---

[13]DeMey, Marc. 1982. *The Cognitive Paradigm*, p. 211. Dordrecht, Holland: D. Reidel Publishing Company.

Automatic behaviors save mental processing time and free up the mind to work on something else. Values and beliefs are not reexamined. This approach represents classic "no need to change what is working," "maintain the status quo" thinking. It is only when a situation is defined as problematic that new information for decision support is sought out. McCown[2] writes that only when a comfortable belief is challenged or it becomes apparent that a high degree of uncertainty or ambiguity is present will a farmer realize the situation is "problematic" and that decisions and actions must be reanalyzed.

Old attitudes and beliefs must be "unfrozen" before being replaced by new ones.[14] Change is not a one-step event "but a complex process in which old attitudes and practices must be open to reexamination (unfrozen) before change can take place."[15]

Conservation Community of Practice

Coughenour's[1] research on adoption of no-till cropping finds that innovative social networks and reconstructed social identities are underlying explanations for whether conservation practices are utilized on farms. He concludes that the adoption of no-till is not individually based but the result of the network that the farmer is connected to. Neighbor-to-neighbor social networks of trust and cooperation are viewed as reliable information sources by members of the network.[16] Information transfer within these horizontal networks of farmers, farm advisors, farm supply agents, and extension agronomists can lead to innovative cropping agriculture based on new technoscience of conservation tillage and new locally developed systems of practice.[1]

The key to adoption of innovation is that farmer traditions of customary practice are commingled with new systems of technical knowledge that are produced by scientists and engineers. As farmers willingly experiment with these new systems, they reconstruct their knowledge base and share what they have learned with other farmers in their network who in turn adapt and modify to meet their own farm system needs.

The community culture of farming practices is negotiated and replicated by other farmers throughout the watershed in everyday practices.[11] It is visible in actual practices in farm fields and reflects what farmers perceive constitute a "good farmer." The visual appearance of the farm conveys a message of stewardship to

[14]Minkler, Meredith and Nina Wallerstein .1999. "Improving Health through Community Organization and Community Building." Chapter 3 in *Community Organizing and Community Building for Health* edited by Meredith Minkler. New Brunswick: Rutgers University Press.

[15]Lewin, Kurt. 1951. *Field Theory in the Social Sciences*. New York: Harper and Row.

[16]Cohen, Jean L. 1999. "American Civil Society Talk." pp. 55–85 in *Civil Society, Democracy and Civic Renewal* edited by R. K. Fullinwider. New York: Rowan & Littlefield Publishers, Inc.

neighbors.[17] When the community culture acknowledges water quality concerns and endorses conservation practices as part of the norm, then the conservation farmer is rewarded. Conversely when the community culture does not recognize water issues or the need to practice conservation measures, then the conservation-minded farmer is not rewarded by the community.

An interview with a northeast Iowa farmer[18] sheds more light on the impact of the dominant practice of widening and straightening creeks and unintentionally making the water go faster thereby changing fish habitat:

> [farmer's name] straightened that one [creek on his farm] out 2 years ago ... it went the full length of his farm. So then the next guy straightens [his] out ... and so on ... took some of the deep pockets out of it for fish.... Nobody really was aware of how much this water was sensitive to what we were doing.... I must be a fool for raising hay, but I do it for conservation reasons ... we're on hilly ground.

This farmer is part of a local network of farmers that are trying to change the community culture by monitoring local water quality to discover how their farm management decisions impact their streams. As they discuss these issues, they learn together, experiment with solutions, and are changing their practices and, as a result, are altering the prevailing community culture to include conservation management. Individual change is possible through social processes that shift the weights of particular values assigned by the community as a whole.[11] The iterative nature of expert-to-farmer and farmer-to-farmer interactions is what makes new learning possible and contributes to the reconstruction of conservation beliefs that lead to new management decisions.

What Are Farmers Thinking? We Need to Ask Them

Voluntary adoption of conservation practices are the centerpiece of restoring and protecting US degraded water bodies polluted by agricultural land uses. Federal, state, and local agencies offer a wide variety of incentives including cost-share programs, low interest loans, and technical guidance. In addition, land-grant university scientists and extension educators provide technological and agronomic support. For widespread adoption of conservation practices within a watershed to occur, we need to understand how farmers think about conservation practices, and their meaning and relevance to the agricultural enterprise.

Conservation innovators don't have a technical script regarding what should be done,[1] but rather develop a technical frame – a set of knowledge, methods, and routines – that can be used to evaluate each field, the season/time/environmental vulnerability, profitability potential in crop selection, and selection of planting and

---

[17]Nassauer, J. I. 1989. "Agricultural Policy and Aesthetic Objectives." *Journal of Soil and Water Conservation* 44(5):384–387.

[18]Farmer #3 NE Iowa interviews 2005.

harvesting technologies. Successful conservation farmers make management choices among crops, tillage, planting and harvesting methods. These decisions can vary from year to year in the same location.

If farmers use a general frame rather than a technical script to make decisions, what are the general themes that frame how they think about available scientific conservation techniques? A survey of 1,110 northwest Iowa farmers[19] asked, "To what extent do you use the following practices to reduce the amount of water pollution (including chemical pollution and soil erosion)?" They were asked to rate 17 conservation practices on the intensity of their use: do not use (1); limited use (2); moderate use (3); and heavy use (4).

Multidimensional analysis[20] of the 17 conservation practices reveal the items cluster into four conceptual groups: record-based management (6 practices, alpha 0.865); field and farm water interception (5 practices, alpha 0.712); tillage practices (4 practices, alpha 0.519); and crop rotation (2 practices, alpha 0.511). Figures 17.1–17.3 show the distribution of the first three groupings. Crop rotation consists of two items, small grain and forage crop rotations, and the distribution is not shown.

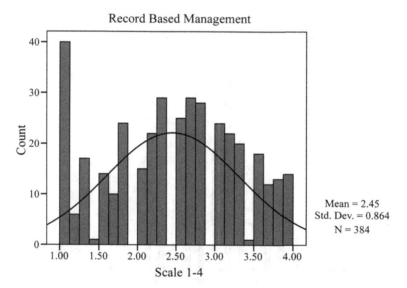

Fig. 17.1 Distribution of record based management scale

---

[19] Lower Big Sioux River Watershed Survey (Iowa) November 2007, mailed to 4,439 farmers, 1,110 completed surveys were returned (25.2% response rate) conducted in 12 subwatersheds of the Lower Big Sioux River by three county Soil and Water Conservation Districts (SWCD). There was a single mailing with no follow-up to increase response rate.

[20] Principal component analysis, varimax rotation; Cronbach alpha test for reliability for each factor. Two factors with an alpha greater than 0.70 have strong reliability; two factors in the 0.50 range do not cluster together as strongly.

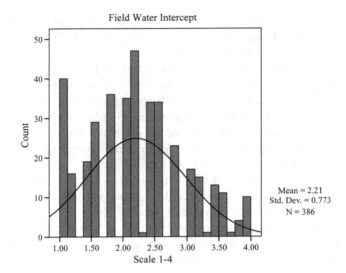

Fig. 17.2 Distribution of field water intercept scale

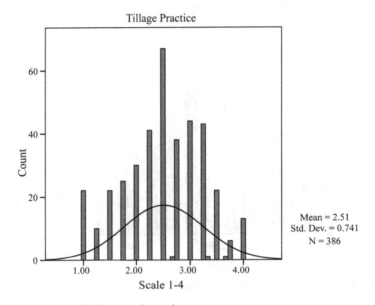

Fig. 17.3 Distribution of tillage practice scale

These groupings give us a sense of the general technical frame these northwest Iowa farmers use when thinking about conservation management.

Six practices cluster under record-based management (Fig. 17.1): soil testing, integrated pest management (IPM), systematic crop scouting, farm-based records, nutrient management, and manure structures. These practices, with the exception of manure structures, have specific record-keeping protocols and entail systematic data collection to document changes over time. When a farmer perceives that patterns are changing out of range of what is considered "normal," it raises the question of whether they should keep doing what they have been or adapt their management decision in response. Routine record keeping provides valuable feedback to the farmer regarding their current automatic management practices and whether the practices need to be revised and adjusted.

The most common management practice in this group is soil testing, used by more than 78% of surveyed farmers, followed by IPM (68.4%), and systematic crop scouting (68.3). The high rates of adoption of practices are likely associated with targeted efforts through extension education programs. Manure structures are correlated to nutrient management and associated with livestock farmers. Farmers under the age of 30 years were most likely to use record-based management.[21] Further, the more acres a farmer farmed, the more likely they were using record-based management. In addition, the use of record-based management was significantly influenced by peer practices.[22] Thus, in this sample, younger farmers, those who farm more acres, and those who interact more with other farmers have adopted this set of record-based practices at a higher rate than other farmers.

Practices that support field and farm water interception (Fig. 17.2) include grass filter strips, grass headlands, grassed waterways, wetland restoration, and enrollment in the conservation reserve program (CRP). Grassed waterways are most heavily used by almost one-third of the farmers. Approximately one-quarter to one-third of the farmers *do not* use some form of field water interceptions: filter strips (32.7%), grass headlands (29.9%), or grassed waterways (23.9%) in this hilly topography. Farmers apparently also think of wetland restoration and CRP in the mix of conservation options they have when concerned about soil erosion and excessive nutrients reaching a nearby stream or river. More than 55% of farmers in this survey report not being enrolled in CRP. The ISU Center for Agricultural and Rural Development reported in 2007 that 34–44% of highly erodible lands (HEL) were in CRP in 2002 in two of the surveyed counties and 74–80% HEL were in CRP in the third county.[23] There were no significant differences in field and farm water interception practices by age.

Tillage practices farmers consider similar as options for reducing water pollution are terraces, reduced tillage, no-till, and contour strip farming (Fig. 17.3). Terraces (48%) and reduced tillage (41.5%) are the most heavily used practices followed by moderate use by many farmers (20.1 and 38.6%, respectively). Almost 55% do not

---

[21] Younger than age 30 years had a mean of 3.10 (Moderate use) and were significantly more likely to use record-based management than four other categories of older farmers; ANOVA.

[22] Age, number of acres farmed, and peer practices predicted 20.8% of the variance in adoption of record-based management with age significant at $p < 0.023$ ($B = 1.155$); and acres ($B = 0.233$) and peer practices ($B = 0.323$) significant at $p < 0.01$. OLS Regression.

[23] Secchi, Silvia and Bruce A. Babcock. 2007. "Impact of High Crop Prices on Environmental Quality: A Case of Iowa and the Conservation Reserve Program." Working paper 07-WP447 Center for Agricultural and Rural Development, Ames, IA. (http://www.card.iastate.edu).

use contour strip farming and 47.2% do not use no-till. In some cases, no-till and contours are used in conjunction with terraces. Neither age nor number of acres nor peer practices in multivariate analyses significantly predict adoption of these water pollution-reducing tillage practices.

Small grain and forage crop rotations clustered together in multidimensional analysis but are weakly associated. In this northwest Iowa three-county region, more than 63% do not use small grain rotations and 50.7% do not use forage crop rotations. Although farmers think of these two as somewhat related practices, neither is considered meaningful or very relevant to their own management. Corn and soybean crops dominate this region and it is likely farmers are not revisiting prior crop mix decisions.

These findings offer insight into the mental maps of farmers in northwest Iowa in 2007 and how they think about conservation practices that have the potential to reduce soil erosion and water pollution. Analysis of the record-based management cluster supports Coughnenur's[1] claim that social networks are important to developing a new technical frame of reference and are associated with adoption of some conservation practices. However, peer practices, age, and number of acres are not always significant predictors of all conservation groupings. This finding illustrates that different conservation practices evoke different responses. Topography, soil types, seasonal variations in weather and market conditions, and difference in watershed cultures are other reasons why what farmers are thinking will not always be consistent within states or across regions or across time. Multidimensional analyses of these same 17 conservation practices are likely to result in mental maps that are quite different after a major flood or in a different location, such as Ohio or New York.

Conclusion

People trust only their own understanding of their world as the basis for their actions.[24] Agricultural management decisions are selectively filtered and processed based on internal motivations, environmental knowledge, values and beliefs about how natural systems work, social connections, problem identification, and how routine the decision appears.[3] If we are to facilitate farmer learning and encourage the adoption of conservation practices on behalf of our water resources, we must understand how farmers think and make decisions. To build this knowledge, we must ask farmers what they are thinking and listen to what they tell us. Facilitation with the goal of increasing conservation behaviors may mean finding ways to *disrupt* automatic management decisions and actions so that farmer goals and beliefs are reexamined.

---

[24]de Geus, Arie P. 1994. "Modeling to Predict or to Learn?" Foreword in *Modeling for Learning Organizations* edited by John D. W. Morecroft and John D. Sterman. Portland: Productivity Press.

New value orientations and identities as conservation farmers are products of new social networks and a sequence of trials and evaluations on their own farms.[1] Farmers who practice adaptive management utilize experiences and science and move beyond technical prescriptions for conservation practice. Structured opportunities to iteratively engage other farmers and the scientific and technological community can expose farmers who have not yet adopted conservation goals and methods to alternative and practical conservation measures available to them. The goal is to help farmers reconstruct their general frame of reference to include ecosystem management considerations in addition to their customary agricultural production practices.

Chapter 18
Sustainability of Environmental Management – The Role of Technical Assistance as an Educational Program

Susan S. Brown and Chad Ingels

Introduction

Technical specialists' responsiveness, listening, and willingness to learn from the farmer builds trust and can increase farmer requests for assistance with environmental best management practices (BMPs). Technical specialists can initiate a collaborative planning process that goes beyond delivering an expert prescription for BMPs to help farmers explore the scientific basis for recommended management strategies and technical options. This process educates the farmer to observe the environmental and profitability outcomes of their decisions and adapt management practices accordingly. We have found that working with individuals in this way, as part of a locally directed performance-based management incentive program, also has a profound impact at the community or watershed level. In this chapter, we show how an extension technical specialist can combine the delivery of expert information with an educational process and bring groups of farmers and landowners into a watershed *management partnership* that leads to more sustainable environmental outcomes.

Technical assistance is delivered as an educational program when specialists go beyond expert prescriptions to become partners who help farm operators understand and explore adaptive management options. The roles of technical specialist and educator are combined by involving the client in participatory learning experiences – listening first and providing scientific and technical details as questions surface from discussion by groups and individuals. This approach is being implemented in a number of Iowa watersheds by Iowa State University Extension (ISUE). The goal of extension involvement is to improve watershed outcomes through adult education and building community capacity by group facilitation and leadership development.

S.S. Brown (✉)
Department of Sociology, Iowa State University, 303 East Hall, Ames IA 50011-1070, USA
e-mail: ssbrown@iastate.edu

L.W. Morton and S.S. Brown (eds.), *Pathways for Getting to Better Water Quality: The Citizen Effect*, DOI 10.1007/978-1-4419-7282-8_18,
© Springer Science+Business Media, LLC 2011

Participatory Learning

Within Cooperative Extension, education is defined as learning that results in a *change in behavior* on the part of the learner and includes both the content and the processes used in relating to the learner.[1] This cooperative education requires a two-way effort in which both educators and clients actively participate. As applied to agricultural nonpoint source water quality projects, cooperative education increases farmers' ability to understand and apply research-based information to flexibly manage their operations in response to changes in costs, regulatory requirements, their personal production and environmental goals, and the expectations of their communities. When education and technical services are provided to the same group of farmers both individually and collectively in meetings or workshops, the result is that participants also educate each other in discussing what they have learned.

When public sector programs provide this kind of assistance, client farmers become educated consumers of technical services, creating new opportunities for private sector consultants to continue assisting them with adaptive management. The cooperative discovery process builds trust between the technical specialist and clients. Our projects have demonstrated that the cooperative discovery approach can make an important contribution to improving the environmental performance of agriculture because of its unique ability to influence producers' knowledge and "ownership" of their farm's environmental outcomes.

Managing Agriculture for Environmental Benefits

Nearly 74% of the Iowa landscape, a larger proportion than any other state, is managed for agricultural crop production[2] and more than 90% of Iowa's water contaminants are attributed to agriculture. Recent water quality modeling research by the US Geological Survey indicates that nutrients from Iowa and other Corn Belt states comprise approximately 56% of the total nitrogen (N) and 46% of the total phosphorus (P) loads delivered to the Gulf of Mexico.[3] The educational projects and citizen watershed councils described in this chapter are located in subwatersheds of the 1.2 million-acre Maquoketa River Basin, a tributary to the Upper Mississippi River. Eight segments of the Maquoketa River are on Iowa's 303d list of impaired

[1] Mahan, Russ A. and Stephan R. Bollman. 1968. "Education or Information Giving?" *Journal of Cooperative Extension*, Summer: 100–106.

[2] U.S. Department of Agriculture National Agricultural Statistics Service. 2007 Census of Agriculture, Volume 1 Part 51, page 347. Washington (http://www.agcensus.usda.gov/Publications/2007).

[3] Alexander, Richard B., Richard A. Smith, Gregory E. Schwarz, Elizabeth W. Boyer, Jacqueline V. Nolan, and John W. Brakebill. 2008. "Differences in Phosphorus and Nitrogen Delivery to the Gulf of Mexico from the Mississippi River Basin." *Environmental Science and Technology* 42(3): 822–830, supplemental material pages S22–S25 (http://pubs.acs.org/doi/suppl/10.1021/es0716103).

waters, including segments in all but one of the project watersheds. The project watersheds are small, 25,000–60,000 acres in size with 80–100 farm operations. Their small size is deliberate, to leverage existing relationships and increase neighbor-to-neighbor interaction. Some of the project watersheds are dominated by row crop agriculture and others by intensive livestock production.

Nationally and locally there is a significant public investment in reducing nonpoint source pollution, primarily soil loss and sediment movement, from Midwestern agriculture. The 2007 Census of Agriculture reports $208 million in federal payments to conservation and wetlands programs in Iowa alone,[4] and state conservation programs provide additional funds. As water quality concerns become increasingly focused on nutrient pollution, technical agencies anticipate much higher public costs to provide additional assistance with nutrient management planning. Yet by itself, a plan only documents intentions. Soil erosion and sediment movement are visible. Therefore, to some extent, so are farmers' accomplishments in reducing this form of nonpoint source pollution. The loading and short-term effects of other pollutants, such as nitrogen, are not so evident. The challenge is to help farmers go beyond having a plan on file to being accountable to their neighbors and communities for reducing the environmental impact of nonpoint source pollutants.

Many farmers are ready to take responsibility for reducing agricultural nonpoint source pollution. A random sample mail survey of citizens in Iowa, Nebraska, Kansas, and Missouri revealed that 40% of farmer-respondents agreed that individual citizens "should be most responsible for protecting local water quality," as compared with 8% of urban residents and 16% of rural nonfarm residents.[5] However, farmers often lack the specific knowledge and skills they need to reduce nonpoint pollution on their own lands. Further they are seldom asked or educated to organize and collectively address their shared problem at the watershed level. Existing conservation programs target and recruit farmers one at a time. They rely on technical specialists to develop and coordinate contracts for expert-created plans on individual farms. Even when these plans are implemented, a change in the condition of the water resource often does not happen in the time scale of a public project, so farmers get no feedback about the environmental impact of their efforts. Because of the individual nature of this intervention, there is also no meaningful measure of watershed-wide accomplishment that can be recognized by the community.

---

[4] U.S. Department of Agriculture National Agricultural Statistics Service. 2007 Census of Agriculture, Volume 1 Chapter 1 State Level Data – Iowa, page 15, Table 6. Washington (http://www.agcensus.usda.gov/Publications/2007).

[5] Morton, Lois Wright and Susan Brown. 2007. "Water Issues in the Four State Heartland Region: A Survey of Public Perceptions and Attitudes About Water." The Heartland Regional Water Coordination Initiative Bulletin #SP289, Iowa State University Extension, page 16 (http://www.heartlandwq.iastate.edu).

Facilitating Citizen-Directed Environmental Management

In Iowa, watershed councils of local farmers are now actively implementing locally directed performance-based environmental management incentive projects. ISUE is a principal institutional partner and its role in these and earlier watershed projects from which the performance concept evolved has focused on facilitating citizen-directed environmental efforts conducted by a peer group. These projects have demonstrated that a council of farmers can provide aggressive local leadership for watershed initiatives. They also show that citizens can rapidly learn and incorporate new science-based information into decision making and that individual and community support for environmental management can be changed as a result. The sociological research presented in Chap. 15 of this volume has described the impact of the performance approach from the perspective of the farmers who participated in the pilot performance incentives project in the Hewitt Creek watershed from 2004 to 2008.

The combination of technical assistance and education progressed through several stages to the current emphasis of on-farm environmental performance. The history of ISUE facilitation of watershed groups in Iowa and the process by which the educational approach taken in performance watersheds was developed offers guidance for other technical specialists wishing to adapt the process to their own interventions.

Facilitation of Watershed Councils

The value of providing technical assistance combined with education emerged from a 1998–2003 pilot study in a subwatershed at the Maquoketa River Headwaters. The site was selected for an interagency project to model the process of locally led, place-based environmental management. Partners included state and regional regulatory and technical agencies, university research, and extension. ISUE had two roles in the project. The first was to provide leadership development and community facilitation, helping residents form citizen watershed councils that could effectively partner with agencies and organizations in managing their watershed resources. Extension's second role was to provide education to build the capacity of watershed producers to implement adaptive environmental management to reduce agricultural nonpoint source nutrient pollution. This education was provided in a cooperative program that involved watershed farmers in active learning experiences and decision-making roles.

The Maquoketa Headwaters project was the first time that ISUE outreach specifically partnered social scientists with agricultural educators to move watershed improvement projects forward. The increased emphasis on human dimensions meant that more attention was paid to learning through participatory discovery as well as other strategies to build proactive citizen leadership and influence peer group and community outcomes. Long-time county extension and Soil and Water Conservation District leaders made initial contacts with influential conservation farmers and commodity organization leaders to sponsor a public watershed meeting.

At the end of the event, volunteers were invited to be part of a citizens' watershed council. An extension community development specialist facilitated the public information meeting and monitored monthly council meetings for up to a year until the councils were organized. Being "organized" meant the group elected leadership, set its own goals and agenda, and was able to plan and implement actions independent of the facilitator.

The purpose of a trained facilitator is to make sure that individuals' views on all sides of the issues are heard, to guide productive discussion, and to assist groups in establishing citizen leadership. All of the project watershed councils are currently active, consist primarily of farmers, and have attendance varying from 8 to 25 attendees per month. Council meetings are not held during planting and harvest seasons. In the longest-running watershed projects, these voluntary councils are in their fifth year.

In some ways, the cooperative education approach was also new to many technical specialists. Instead of experts delivering the facts citizen-farmers needed to know, they accepted the role of facilitators and advisors. At watershed council meetings agency staff, other invited experts and visitors were seated around the group while the citizen council occupied a central table. The job of the specialists was to provide science-based information and technical resources as questions were raised by the council, rather than trying to lead the process. Councils saw themselves as gathering information about local environmental issues on their own initiative, and were found to quickly grasp the basic science behind most water quality BMPs. They also rapidly identified practices judged to fit best with their current operations and became interested in testing new ideas. John Rodecap, Extension Coordinator for the Northeast Iowa watersheds from 1990 to 2008, noted that when farmer councils were able to lead their own discussions, they "often started out in a direction you didn't think was right (as an educator, on the issues at hand) but quickly came around to where you wanted them to be" in dealing with facts about pollution and realistic ways to manage it.

Making Data Accessible

Monitoring

What citizens almost never quickly accept is the externally imposed impaired waters designation. An important motivator for many watershed leaders is to do something preemptive about pending regulatory action, to "get our stream off the 303d list." However, first they have to accept the assessment. The importance of participatory discovery is shown by the fact that in every watershed citizen councils have decided to sponsor volunteer water monitoring and conduct local surveys before they will accept some of the nonpoint source assessment information coming to them from state and federal agencies. Once they have accepted it, they are ready to seriously consider solutions and act to reduce their potential contribution to the impairment.

Chemical and biological monitoring in project watersheds has variously been with assistance from the state IOWATER citizen monitoring program, council members,

FFA students, high school and college science classes, or in partnership with other monitoring programs. An important contribution of extension to this local monitoring has been making the information contained in the raw data understandable and facilitating discussion of results among local citizens. Ideally this process involves presenting, summarizing, and helping people interpret the data. Figure 18.1 shows an example of how monitoring results were tabulated with incremental additions *and significant items highlighted* for the North Fork Headwaters monthly council meetings. The first item of business at every meeting is to review these results. Repeated discussion educates participants about water measurement and the impact of land management and weather, about data quality issues, about pollution problems they can or cannot solve. Most important it orients the rest of the meeting and leads to more open discussion of possible nonpoint source problem areas in the watershed.

Note that the figures in this chapter are intended as examples of how educators have summarized and organized technical information in graphical and tabular form for discussion by local groups. They are effective because the information is watershed specific and also because economics and practical details are always included. Examples relevant to local situations and known council priorities are always included.

Modeling

In the first Maquoketa subwatershed projects, Extension worked in partnership with research scientists from the Texas Institute of Applied Environmental Research and the Center for Agricultural and Rural Development and Department of Agricultural and Biological Engineering at Iowa State University (ISU) who conducted monitoring and developed models to assess the environmental and economic impacts of alternative crop, livestock, and land management practices. In typical impaired watershed situations, modeling results are incorporated into a total maximum daily load (TMDL) and the watershed management plan written by an expert with recommendations "delivered" to local producers. In the Northeast Iowa projects, the farmer council was directly engaged in the research and planning in two ways. First, they provided detailed input on local crop and livestock management practices to refine the model baseline assumptions. They were allowed to request specific management scenarios they wanted to explore. The council put forth a number of scenarios and dropped some that were proposed by the modelers and extension specialists because they were convinced those scenarios would not be readily adopted in their watershed. The application of models proved to be an interactive process among the modeling scientists, extension staff, and members of the council. The scientists involved said the level of cooperation with local land managers was unique in their modeling work.

Second, the council was given a chance to combine their local knowledge with science-based information by recommending priority cost-share practices for a state-funded conservation project. The extension specialist initially summarized and simplified the modeling results into graphical pie charts from which the council could quickly grasp the relative effectiveness and costs of specific BMPs for pollutant reduction. Figure 18.2 is an example. These charts were used repeatedly with the

North Fork Maquoketa Headwaters Monitoring, 2009
Rick Klann, Upper Iowa University, Fayette, IA

Site 1

Date (month/day)	3/18 7/22	4/1 8/8	4/27 9/7	5/8 10/2	5/14	5/28	6/8	6/30
24-hour rainfall (inches)	2.1	1.1	>1	0.8	1	rain*	2.5	
Water temperature (°F)	50 / 66	44 / 67	53 / 64	58 / 55	54	56	60	65
pH (acid, <7.0 normal > basic)	8.2 / 7.8	7.2 / 8.3	6.9 / 8.4	6.8 / *	7.2 / **	**	**	8.5
Conductivity (OK 400-1000) (ions, minerals, salts)	643 / 619	622 / 918	393 / 668	620 / 664	457	645	342	673
Dissolved oxygen (mg/l)	9.9 / 5.8	12.5 / 5.5	9.5 / 11.4	13.1 / 8.9	9.6	8.1	7.9	8.1
Turbidity (water clarity)	23 / 51	24 / 26	>1,000 / 9	10 / 36	>1,000	39	2,685	33
Suspended solids (mg/l) (<20 good)	35 / 50.7	29 / 51.3	420 / 9	14.2 / 36	532	53.6	1,860	52
Total phosphorus (mg/l)	0.74 / 3.56	1.76 / 2.79	2.89 / 0.28	0.41 / 1.03	4.7	1.94	6.43	0.86
Total nitrogen (mg/l)	6.6 / 9.4	6.6 / 8.6	12.8 / 5.6	6.4 / 5.4	16.7	10.3	18.8	6
Fecal coliform (cfu/100 ml)	80 / 1,329,000	560 / 381,000	29,000 / 1,940	30 / 39,000	112,000	302,000	246,000	4,680

Fig. 18.1 Monitoring data for watershed councils. Review of updated monitoring results at the beginning of each watershed council meeting keeps participants focused on environmental issues

Nitrogen and Phosphorus Loss Reduction

Full nitrogen accounting from manure to reduce commercial N use

43% less soluble N, saving $9.24/A

Exclude crop removal fertilizer applications on high testing fields

12% less soluble N
32% less soluble P, saving $18.86/A

Manure application at the phosphorus crop removal rate

31% less soluble N
22% less soluble P, saving $7.14/A

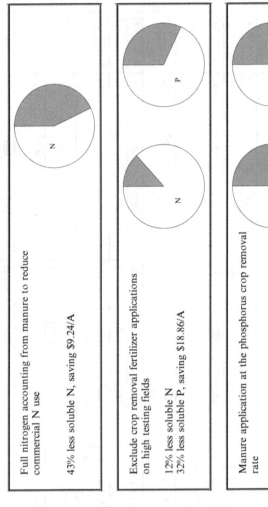

Fig. 18.2 Environmental cost-effectiveness scenarios for N and P loss reduction. Environmental cost and effectiveness of management scenarios as predicted by watershed models are put into graphic form for educational programs

council to focus project direction. Unlike research presentations, the farmers do not need a lot of scientific detail to get the idea. And they accept the results because they had input into the modeling process itself.

The response of producers who participated in the modeling was marked. The modeling resulted in detailed scenarios with flexible options that could be adopted to maximize the environmental benefit available to individual operators depending on their current financial and management resources. Having options for adaptive management and research-based information on environmental efficacy motivated many producers to go beyond what public programs would have required. Significantly, 5 years after the modeling activities, members of these watershed groups still kept and referred to the cost-benefit charts and descriptions in discussing their environmental management options.

On-Farm Demonstrations

Adoption-diffusion studies have consistently shown that producers accept new management practices slowly, wishing to evaluate results over multiple crop seasons. The economic and environmental benefits of nutrient and manure management BMPs are often emphasized in watershed projects with research and cost-effectiveness modeling results. However local demonstrations, conducted close to home, offer producers proof on their own farms that practices such as improved manure management are beneficial. Extension has found that once farmers learn their experiences and knowledge are valued in group discussions, they not only are willing to host on-farm demonstrations to test practices, but will initiate the questions and jointly explore finding effective solutions.

Extension technical specialists who assist the watershed on-farm demonstrations have recognized the teachable moment and have made education and responsiveness a high priority. Each demonstration site includes a treatment replicating the cooperator's typical practice. A scientist conducting on-farm research may not consider such a treatment a priority, but farmer cooperators are much more likely to go out of their way to manage demonstrations carefully and discuss the process and results with their neighbors if the demonstration addresses their specific concerns. Plot results are delivered to the cooperator the day they are collected and local summaries are made available to cooperators before they are publicized in the community. Prompt feedback shows respect for cooperators and encourages them to include demonstrated BMPs into their fall preparation for the following crop year. Cooperators who are confident they understand their results become advocates for the practice and will often ask, "What else can I do?" In crop years 2000–2003, as a result of the council discussions, producers in three Maquoketa subwatersheds voluntarily hosted more than 50 on-farm demonstrations of manure, fertilizer, and tillage management strategies to control P and N loading. Extension has conducted similar demonstrations in many Northeast Iowa watersheds.

Figure 18.3 is an example of how demonstration results are presented to watershed councils and in local publications. They provide solid evidence that reduced fertilizer

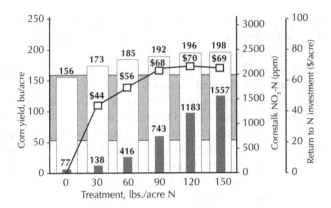

Corn yield, end-of-season cornstalk nitrate-N results and return to N investment
from 19 small-plot N rate demonstrations on corn-following-soybeans, 2000-03.
(Corn at $2.40/bushel and N at $0.20/lb.) The shaded rectangle in the background
of the graph indicates the optimal end-of-season cornstalk nitrate range of 700
to 2,000 ppm.

Fig. 18.3 On-farm nitrogen rate demonstrations. Four-year summary of on-farm nitrogen rate demonstrations in Northeast Iowa. Farmers like having many elements of their decision making displayed together

application and improved manure management are profitable, and they are most believable because they are conducted on-farm with rates and treatments similar to what local managers use. The multiaxis plot includes multiple items needed for adaptive management – rates, yields, costs, and performance measures. Farmers quickly get comfortable studying information in this format.

Small-plot and field demonstrations initiated by farmers, with design and implementation assisted by specialists, can balance cooperators' interests and capabilities with replicated treatments and measurement precision. Field-scale, technology-assisted demonstrations are unique because farmers see the results in real time and report their interpretations to the project coordinator, rather than just receiving a paper report. Providing technical assistance as an education program also means that data collection and presentation from plot results should go beyond yield, to introduce management tools such as manure spreader calibration, evaluation of soil capability and soil test results, diagnostic tests for late-season stalk N in corn, and economic returns.

Technical Assistance as Education

Incentive Education Program

In a separate project, a nutrient management incentive education program was developed for nutrient-impaired watersheds. Farms in many of the Northeast Iowa watersheds are livestock intensive with a small land base. These potentially high-environmental impact operations can profitably use on-farm nutrients, but the technical service-provider model for nutrient management plan (NMP) implementation is difficult to apply. Given a choice, crop consultants prefer to serve large-acreage corn–bean rotation farms where management issues are less complex and they can receive more income from their time. Sustainable change in nutrient loading in this landscape requires producers to take a more active role in refining nutrient and manure management on their own farms.

The model for the nutrient management incentive program was producer education rather than delivery of management prescriptions. The program was designed to foster the process of discovery, through which participants learn to plan, evaluate, and progressively refine *their own* site-specific nutrient management under the guidance of a university extension field specialist. Rather than being led through sample exercises, participants bring in their own records and the record keeping required for the program includes evaluation of economic outcomes. A total of approximately 30 farmers participated in the first program and received a small per-acre payment for fields enrolled. Workshops had 8–12 participants. The small group made efficient use of the specialist's time, while still allowing for individualized assistance. Workshops were tightly focused on soil agronomic potential and optimizing use of manure and legumes as on-farm nutrient resources, with supplemental commercial nutrient application, for profitable crop production.

Participants attended one or two workshops each spring and fall for 2–3 years. The multiyear series gave them time to evaluate management changes over several crop seasons. As the program progressed, participants began implementing refined practices on an increasing number of fields. The workshops also captured the reinforcement provided by participant comments, group discussion, and observing peers. Following workshop discussions, farmers had more focused questions for the specialist when consulting individually because they had background information and were ready to try something specific. Demonstration results were shared with the group and cooperators often spoke to others about specific results from their demonstration sites. During later sessions, producers would ask questions of the ISU specialists and their neighbors would be the first ones to answer, certainly giving more credit to the responses.

Demonstrations of manure nutrients were a primary educational component. Priority fields were defined from soil test data and focused on the economic benefit of reducing phosphorus application rates where soil test levels were high or very high using University recommendations. Participants found that their agricultural suppliers did not necessarily use university recommendations when providing field-level nutrient recommendations. Even with detailed grid sampling results that cost $8–10 per acre, the supplier recommendations might still be over-applying phosphorus.

Asking for soil test results as part of the technical assistance forced the producers to get the information in their own hands from the supplier, or at least take the time to review the data they already had accessible at home. Many farmers had never taken time to study their soil tests, relying on their fertilizer dealer to make an application recommendation. Once the information was in their hands, most made some sort of change, whether it was the elimination of crop removal fertilizer applications on high-testing soils or moving manure to fields or field areas where it would be most beneficial.

This education model for NMP implementation helped put responsibility for watershed nutrient management directly into the hands of watershed farmers. Participants realized savings from fertilizer not applied of up to $2,800 per farm and their comments showed they were taking increased credit for manure nutrients and implementing practices such as managing fertility by soil map unit (and thus, potential yield) on more acres. However, while the quantity of product not used is important, it is less so than the educational outcome – that participants understand and accept the technical basis of the new plan and are motivated to apply it in future revisions of the plan. Therefore, the incentive education program measured accomplishments primarily as changes in participants' knowledge, attitudes, and management skills.

Between preproject and postproject surveys, there was a significant increase in the number of participants who responded to the question "who is primarily responsible for your crop fertility program?" with "I am." This does not mean they were not still using professional services of their suppliers or crop consultants, but the program gave them confidence in their ability to ask the right questions of suppliers. They had assumed increased responsibility for nutrient and manure utilization decisions, and also for nonpoint source pollution reduction, on their own farms.

Performance Measures for Environmental Management

The nutrient management incentive education program gave farmers skills and knowledge to practice better environmental management using data from their own farms. In this process, the participants also become familiar with agronomic planning tools – models and field tests – some of which are aimed at environmental as well as production issues and can be used to assess environmental performance. Soil testing is a well-established production performance measure, but not an environmental performance measure. Its results may allow farmers to significantly reduce their P application, but it does not make the connection between reduced application and improved water quality. There are measures, however, such as the Iowa Phosphorus Index (P Index), that are based on environmental models, and their results make an explicit connection between field-level practices and impact on environmental quality.

ISU Extension is currently facilitating councils and providing technical and educational assistance to citizen-directed on-farm environmental performance incentive programs in four Northeast Iowa watersheds. Performance measures that local farmers have learned to use and evaluate are the P Index, the soil conditioning index (SCI), and the late-season corn stalk nitrate test (CNT). All of these measures are agronomic decision tools familiar to agencies and crop specialists if not to farmers themselves. The SCI is a model-based predictive tool used by Natural Resource Conservation Service (NRCS) in conservation planning to estimate whether applied conservation practices will result in maintained or increased levels of soil organic matter. Soil organic matter is strongly affected by management practices, and improves the soil's capacity for nutrient cycling, filtering and buffering of pollutants, and resistance to erosion. The SCI does not indicate a desirable or target level of soil organic matter, but it will predict if a particular management system will have a positive or negative effect on soil organic matter. Therefore, the SCI is recalculated for a field from year to year as new practices are applied.

The P Index is required as part of a NMP according to the EPA technical CAFO guidance. The P Index is a simple model-based assessment tool that evaluates landscapes and management practices for potential risk of P movement to water bodies through the identification of critical P source areas. It was developed as a qualitative tool to rank site vulnerability to P loss. The P Index can also be used to help identify alternative management options and provides flexibility in developing remedial strategies. The annual recalculation of the P Index on a given field can measure qualitative improvement (performance) over time when environmental P management practices are implemented.

The CNT is recommended by many land-grant extension programs as a tool for evaluating and refining nitrogen fertilizer management. Corn stalk samples are taken after the grain is physiologically mature and the results evaluate the adequacy of the producers' nitrogen program for the current growing season. Plants that have had excessive amounts of soil nitrogen available (more than is needed for maximum yields) will store higher concentrations of nitrate in the lower stalk sections by the end of the growing season. Although the test is postmortem for a given growing season, it gives producers an opportunity to evaluate their N management over the long term and refine/reduce excess N application and environmental loading. As applied in the incentive program, the test provides a "report card" for participants to evaluate their current N fertilization practice, and feedback that may result in reduced application the following crop year.

On-Farm Performance-based Management Incentives

The pilot performance-based management projects in Northeast Iowa watersheds arose from the expectation that farmers would voluntarily adopt practices to meet watershed and personal environmental goals if they had a way to assess environmental progress on their own farms, just as they currently have measures to assess their crop and livestock production. The unique aspect of a performance-based incentive program is that producers are rewarded for improving the values of their environmental indexes (predicted water outcomes) by whatever practices they choose to adopt, rather than receiving cost share for specific practices. The program gives cooperators flexibility to consider BMP options that fit in with their current operations, and economic and management resources. The index models are also tools they can use to practice adaptive management and continue to improve in the future. Having measures of progress has been shown to have just this effect, motivating producers to continue adopting further changes and target problem sites.

The first performance-based environmental management incentive program was in the Hewitt Creek Watershed. Local leaders began with a goal of getting their creek off the Iowa impaired waters list. They obtained funding for incentives from the Iowa Farm Bureau and the Iowa Corn Growers, and subsequently from the Iowa Watershed Improvement Review Board (WIRB). Extension facilitated council formation and provided education and technical assistance with support from the USDA NIFA National Integrated Water Program, EPA Region 7, and the Iowa WIRB.

Education for Watershed Councils

The Hewitt Creek watershed council (and later Lime Creek, Coldwater–Palmer, and North Fork Headwaters councils) developed their own incentive program, starting with a simple template based on the three measures described above. At the outset of a project, extension found that individuals do not easily connect agronomic management to water quality. Many times the stream is not close to their property, their fields do not drain directly into the stream, and they have never given much thought about how their management might affect the water and environment past their field boundary. Water monitoring and a collective discussion about the results bring the importance of water quality improvement into focus. Making water quality improvement a priority begins at the watershed council. Their monthly meetings always start by reviewing monitoring data and how it might relate back to the farm and field level.

Councils, which manage performance incentives in their own watersheds, are open to anyone in the watershed who wants to participate, as is their incentive program. To develop the incentive program, the councils must understand the performance measures and then learn to compare management practices that might impact their results. Extension provides watershed-appropriate management scenarios for the group to use as a starting point. It is important that these scenarios use watershed-specific data and include scenarios requested by the group. This menu of scenarios may expand from year to year as their interest in performance expands.

Extension also facilitates council discussion, in general terms, of which practices are most acceptable in their watershed and how farmers might adapt management to improve performance. From this discussion and their study of watershed impairments, the councils set and annually review priority practices, performance results, and incentive payment levels. Decisions are based on consensus and members take a vote if needed.

Extension specialists find that educating the councils about the performance measures during the incentive development process brings the work of the watershed council and education of the individuals much closer. In each of the watersheds, everyone involved in the council participates in the incentive program and can explain to their neighbors what practices affect the performance measures.

Technical Assistance as Education

Once the performance program is started, extension specialists also meet individually with cooperators, calculating the performance indexes for each of their fields. A whole farm-weighted average is the number used to determine whether performance has improved enough to earn the incentive "reward" each year. Results from all individual fields in the watershed are also tabulated and reported to the group. Participants can evaluate the number of problem fields and track progress at an individual and watershed level. The tables list index values from worst to best and colors indicate breaks in the data between high, medium, and low Index results, or excess, optimal, and insufficient, in the case of CNT results. A summary of watershed averages is also included. Field designations are set by each participant and farm identifications are anonymous to others, however each cooperator knows which fields are theirs. Over time the groups become very engaged in their progress on their own farms and compared with others. They become willing to identify themselves and share specific information with others.

After calculating indexes for their fields, the extension specialist discusses management scenarios that the cooperator can consider to improve their indexes in the following year. These are suggestions only, not requirements to remain in the program. They are farm-specific versions of the general management scenarios developed for the councils. Figure 18.4 is an example of a set of scenarios. Having these options laid out creates many individual questions about practice implementation for priority areas on cooperator farms. "How might a buffer along the creek in this field work?" "What would no-tilling corn into sod do for my P Index?" Because each cooperator already has their own farm data, they do not ask "Where do I start?" because they already know which fields should be addressed first.

Cooperators receive individualized written reports about their performance levels in a tabular form and also spatially on a farm map generated from the most recent aerial imagery. In Iowa, annual aerial photography has been available since 2004. The map is also used during the next year's review session so that the farmer and specialist are talking about the same fields. The farmers like the aerial photos and ask for additional copies. The report also includes a short handwritten note that explains their incentive payment. There is a small incentive payment to farmers who conduct annual reviews with the extension specialist.

Farmers will make a change if it makes sense and works into what they are doing. Generally small changes are made each year. For example, one farmer tried no-till planting on two fields 1 year, he had good results and the next year he no-till planted his entire farm.

Performance management scenarios for row crop production in Hewitt Creek Watershed				Continuous Corn (CC) Fall chisel, field cultivate, annual manure		CC w/field buffer 50' field-edge vegetative buffer		Corn/Soybean Rotation Fall chisel after corn, field cultivate, manure		2 Corn, Oat, 3 Alfalfa Fall chisel, field cultivate, manure before corn	
Field ID	Soil Map Unit	Acres	Soil P	P Index	SCI	P Index	SCI	P Index	SCI	P Index	SCI
3	482C	32.2	130	4.24	0.57	3.52	0.57	4.81	0.28	2.65	0.78
H2	163D2	27.7	54	5.05	0.21	3.97	0.21	5.52	-0.06	2.72	0.61
6	162D2	31.7	52	3.13	0.35	2.50	0.35	3.43	0.14	1.80	0.71
1	394B	35.2	105	2.17	0.62	1.90	0.62	2.43	0.43	1.58	0.84
Total acres =	126.8										
Weighted avg. index values				3.56	0.45	2.91	0.45	3.96	0.21	2.16	0.74

Fig. 18.4 Performance management scenarios. A technical specialist discusses management options rated by their estimated impact on P loss and soil quality. Farmers have flexibility to accomplish their environmental goals

Neighbor-to-Neighbor Learning

The success of working with farmers individually comes from priming the individual through the education of the group. Group education takes place during the process of identifying the problem, developing a course of action (such as an incentive program), and reviewing watershed performance and progress. The discussions within the group setting lead to individuals considering options that were previously not given much thought.

During watershed council discussions, farmers educate each other about how specific practices worked. For example, in Hewitt Creek, several cooperators tried cover crop after harvesting corn silage, some tilled and then planted while others no-till planted, most used rye while some tried oats, some harvested before planting another crop in the spring while others killed the cover crop before spring planting. These farmer-designed comparisons led to refined cover crop management in the watershed.

Reviewing performance results listed by field from worst to best also leads to discussion about how neighbors achieved good test results. Cooperators learn from their neighbors that there might be a different way to manage a field because often their poor-performing fields are near good-performing fields with similar soil that differ only in management.

One item each watershed council has included in their incentive program is an ongoing performance payment for a cooperator's current level of performance. The performance incentive programs have attracted a very high level of participation – up to 70% of watershed operators and a majority of watershed cropland enrolled. Councils believe that neighbor-to-neighbor support is important for the watershed and want to keep all potential cooperators interested, not just those who need to make the most improvements. Cooperators with good performance levels continue to make changes in management or add practices that have further improved their whole farm performance. Some cooperators whose fields have high baseline performance scores also have a "problem" area that is very difficult to address. Even if they work on the problem area these cooperators may have a hard time raising their overall farm average. The ongoing performance payment keeps this group engaged in the process as well.

The councils also offer a small incentive payment for watershed-wide participation and water quality improvement. The extension specialist calculates the percentage of residents cooperating and percentage watershed area of fields enrolled. This approach draws participants into a watershed community and encourages members of the group to recruit new cooperators. The incentive based on water quality improvement focuses the group on nonpoint source improvement as a project they are all in together and gives the council a way to recognize the group for collective achievements.

Hewitt Creek Accomplishments

In Hewitt Creek, an extension technical specialist has assisted in tracking performance indexes on 396 fields totaling 9,893 acres on 47 farms. High environmental-risk fields with P Index values greater than 5 received the most attention by cooperators, resulting in 39% improvement in P Index scores. SCI values on these priority fields improved 91%. Watershed-wide P Index values on 3-year cooperator farms improved 14% and SCI scores improved 10%. Index scores improved most dramatically through the implementation of no-till planting, new waterways, contouring, and cover crops.

Using the Sediment Delivery Calculator, sediment delivery to Hewitt Creek was reduced by 1,366 T/year and P delivery by 1,776 lb/year during 2008, bringing the project annual sediment reduction to 4,033 T sediment/year. Cornstalk nitrate samples were analyzed for 36 farms, showing a residual nitrate reduction of 51%. Due to cornstalk nitrate results, project cooperators reduced nitrogen applications by an average of 44 lb/acre, a 22% reduction that impacts 8,537 acres and would save 220 tons of nitrogen annually, if extended to all watershed acres.[6]

These performance results have shown up in water monitoring data with a general trend of improved nutrient and suspended solids analyses and improved late-summer dissolved oxygen levels. Improved diversity and quantity of macroinvertebrates were also found during semiannual evaluations.

Comment

Conservation management is just one of the pieces of information farmers are bombarded with on a daily basis. Others such as crop yield, rate of gain, prices of agricultural inputs, and grain and livestock markets are used by farmers to educate themselves prior to making management decisions impacting their operations' bottom line. An environmental performance measure can be another tool that farmers use to adjust their management, but, to be effective, a performance incentive needs to be relevant at the farm level, be watershed specific, and connect the watershed-level pollutants to field-level management. The performance measure should help the farmer set priorities and gauge improvement, and allow for flexibility and innovation. Technical assistance provided as an educational program to groups and individuals can deliver these tools. The locally directed performance approach has been shown to bring about significant change in farmer knowledge, motivation, and environmental practices that benefit their farms, their communities, and their watersheds.

---

[6]Rodecap, John. 2008. Hewitt Creek Iowa Watershed Improvement Review Board Progress Report. Progress report to the Iowa Watershed Improvement Review Board, Iowa Department of Agriculture and Land Stewardship, Des Moines IA.

Chapter 19
Building Citizen Capacity

Susan S. Brown

This book represents the experience and knowledge of researchers and practitioners who study and work with people and communities to restore and improve water quality. It is intended to offer insights and flexible guidance to watershed leaders and specialists as they develop their own approaches to draw out the citizen effect for improving watershed management. The citizen effect refers to the many ways people engage science, technology, and each other to identify and solve their watershed problems. These interactions are not well known to many natural resource professionals, including extension agronomists and agricultural engineers who are often asked to provide watershed management education. In 2005, the Heartland Regional Water Coordination Initiative hosted a Regional Water Conference for an audience made up of educators, technical specialists, and community development practitioners – individuals who work at "ground level" in watersheds. Some of the attendees were already associated with nonpoint source watershed projects as project coordinators or were providing assistance and sponsorship through their county and agency organizations. Others were hoping to initiate watershed work. A postconference survey found that, of all of the technical sessions offering information on regional priority water issues, those related to citizen involvement in watershed management were of greatest interest and perceived value to attendees.

Many of the participants in the Heartland Conference had already come up against two of the most problematic aspects of water quality projects. The first is that nonpoint source pollution management needs the voluntary participation of many individual landusers to be effective, but there is no broad agreement among rural residents that water quality is a problem. The second is that both public and private funding sources for watershed planning and implementation increasingly require projects to design in some form of public participation, but there are few models for how to do this. Further, most of these specialists are not trained to provide group process facilitation and social interventions.

S.S. Brown (✉)
Department of Sociology, Iowa State University, 303 East Hall, Ames IA 50011-1070, USA
e-mail: ssbrown@iastate.edu

L.W. Morton and S.S. Brown (eds.), *Pathways for Getting to Better Water Quality: The Citizen Effect*, DOI 10.1007/978-1-4419-7282-8_19,
© Springer Science+Business Media, LLC 2011

Because the Heartland Initiative proposed citizen involvement as a regional priority water quality issue in 2003, the team has consistently received feedback from small and large regional events asking for case studies. Yet the majority of published papers and reports on citizen/community involvement in watersheds have been case studies. What are these professionals missing that they feel they need more? A likely interpretation is that everyone who works at the local level understands well that every case is different. Every watershed and watershed community has its own culture, norms, patterns of conflict and power, and environmental problems. When case study reports do not analyze their results in a larger theoretical framework, it is easy for the outreach specialist to think "but that wouldn't work in my watershed/community." On the other hand, when presented with theoretical treatments that are not generated from watershed work, such as the "theory of reasoned action" – a currently dominant theory of decision making – people in the field quickly decide that their actual experience does not match the results predicted by the theory.

The field of sociology has developed practical knowledge on "how" to involve citizens in community development and has generated standard methods for organizing local actions. These methods have been adopted into a number of manuals currently available on how to get citizen participation in watershed management. However, such manuals also tend to assume that citizen participation is important, rather than making a cogent argument for why and under what conditions citizen involvement is most useful. And they can give the impression that citizen participation is something that can be accomplished with a checklist rather than giving perspectives on how to obtain citizen engagement in the context of power relationships, cultural conflicts, or other human dimensions issues.

Field specialists are on the front lines. What are they really asking for when they ask for case studies? They work with communities, with farmers, with watershed groups. Citizen participation is mandated. Most are not trained to deal with it. We believe they are really seeking three kinds of information, presented at a theoretical level, to incorporate into their professional practice. These are: (1) how to get citizens involved – why and where the various successful strategies work, (2) how to measure successful citizen participation, both to improve their efforts and to be accountable to funders that expect public involvement, and (3) what can citizen participation really accomplish and how does it complement the technical approaches already available? The chapters in this book present both theoretical and applied treatments of these three aspects.

How and When to Get Citizens Involved

Just as field specialists and educators can range widely in the extent of their training for facilitating the human aspect of social organizations and local decision making, they can also vary widely in their motivation for trying to involve citizens in watershed management. Some will be at one end of the spectrum, regarding citizen involvement as a time-consuming agency mandate that will not improve the final environmental outcomes in their watersheds, or as a way to reduce public opposition to expert decisions that have already been made. Some will be at the other end of the spectrum, seeing citizen education and leadership, whatever its difficulties, as a moral imperative for public programs and management of public resources in a democratic society. Most people feel the need for giving citizens a role lies somewhere between these extremes. They recognize that citizen participation in getting to better watershed outcomes can depend on the situation. As noted above, there are various handbooks on process – but there is a lack of research-based knowledge that connects social science research to practice in determining and implementing appropriate citizen involvement.

One objective of this book is to offer successful research-based strategies for bringing citizens into the process of watershed management in a meaningful, substantive role. Our authors discuss what is known about people's thought processes and motivations and identify the key elements of social networks and interactions that can be used to engage and motivate individuals and communities in watershed decision making.

One important factor in watershed management is regulatory pressure. As was found in Hewitt Creek Watershed, this can effectively energize a proactive local movement (Morton and McGuire, Chap. 15). Conversely, in the eastern Nebraska saline wetlands, invocation of a regulatory program stalled out local action (Corey and Morton, Chap. 13). The difference between these two experiences lies, in part, in the extent of local ownership of the response. One watershed had more community members present in leadership roles, the regulatory pressure was mainly from the state rather than the federal level, and citizens perceived that enforcement could include flexibility for place-based solutions. In the other, a federal law (the Endangered Species Act) blocked local capacities to integrate environmental protection goals with landowners' expectations for their lands.

Another factor that influences success is the need to work across political boundaries. The impact of this factor is related to the size of the targeted watershed. While the watershed has been recognized for many years as a "natural" resource management unit, funding is often directed to large watersheds that span hundreds of square miles and multiple political units – villages, counties, urban centers, and even states. Depending on the stakeholders, involving citizens across political boundaries can be virtually impossible in these situations. There are too many sources of conflict and too much competition for resources among different political entities to leave room for grassroots programs. Effective strategies in large watersheds have targeted political units and large-scale voluntary organizations rather than citizens themselves, and have facilitated ways for these entities to work together in policy networks (Selfa and Becerra, Chap. 10) or as communities of interest (Pfeffer and Wagenet, Chap. 9).

Other projects described in these chapters, notably the Pennsylvania Community Watershed Organizations (Brasier et al., Chap. 11) and the Ohio Sugar Creek Partners (Weaver, Moore, and Parker, Chap. 16) deliberately focused their resources on small watersheds to increase their chance of success in mobilizing community support and peer pressure. In these watersheds, the water resource concern touched enough residents directly to draw out and engage their participation. However, even when not directly affected, residents in a smaller watershed have increased contact with each other in towns, school districts, churches, and other community organizations. These neighbor-to-neighbor connections can offer valuable support to a water project. As Morton's research demonstrated (Chap. 4), social networks have a significant influence on farmers' concern about water resource problems and motivation to improve their own conservation practices. Projects in smaller watersheds also report that neighbors recruiting neighbors is more effective in building local participation than the efforts of an agency specialist. It appears the effect is more a matter of trust and established communication than of agreement on the issues. It can take several years for watershed project staff – "the government" – to gain the trust of watershed residents. This is also an argument for keeping senior local agency staff closely involved with watershed projects, to take advantage of the trust already built between themselves and the community.

Another factor to be taken into consideration is the diversity of cultural norms, or lack of diversity, among groups of local citizens and other watershed stakeholders. Negotiation of cultural differences and building trust among watershed stakeholder groups to pursue a common goal – especially at the outset when the goal is initiating dialog – can be a key element in achieving meaningful citizen participation. The frequent recommendation that watershed groups must be inclusive and reflect the full breadth of diversity in watershed interests may not actually be the best place to start. Work in Sugar Creek (Chap. 16) with different religious groups, separate agriculture and urban initiatives in the New York City watershed (Chap. 9), and trust-building projects with Native American communities (Chap. 14) are all examples of allowing culture-based groups to cultivate their own watershed goals or vision prior to bringing them together with those of differing interests. Zacharakis (Chap. 5) describes conflicts arising among different cultures as well as among political and economic interests, including the difficulty agency representatives find in communicating with residents. As an outside facilitator, he exploited this conflict to encourage residents in developing their own vision for a local conservation program in preparation to work with agency projects.

Comito and Helmers (Chap. 6) find that cultural differences among agencies can also create conflict and frustration. Agency cultural differences arise from differences in their mandates – education, technical assistance, and regulation. Comito and Helmers' "re-languaging" concept challenges us to think anew about the words we use to discuss agricultural practices and the environment. They propose developing a shared "culture of conservation" through new language about impairments and practices and new definitions of the good farmer. Such "re-languaging" is an important function of local leadership, in both the public and private sector, as the community deals with contentious watershed issues. Leaders create structured opportunities for talking about water concerns and guide productive discussion among citizens and groups so that areas of agreement and disagreement are transparent but mutually respected. This lays the groundwork for trust and leads to the ultimate goal, shared action. A culture of conservation is a culture of communication.

Measuring Success in Getting Citizens Involved

The problem of documenting and evaluating citizen involvement is a distinct issue from the practice of engaging citizens so they act collectively to improve their water resources. Our research identifies several core themes in the practice of citizen involvement that must be addressed so that accepted community development methods of group and leadership facilitation can be put to work. These include how to recognize and respect individual and cultural differences, when to engage the process of active learning, and challenges in navigating political and economic conflict. There are both common patterns as well as temporal and contextual factors that make each watershed unique. At a given time in a given watershed, the type and depth of public participation needed depends to some extent on what kind of change has to happen and what part citizens can most effectively play. It is as much art as science to guide a community through the awareness, learning, goal setting, actions, and monitoring processes. However, if we are to make effective use of citizen involvement, it must be documented and evaluated. This is critical not only from funding and accountability standpoints but also to refine and guide the social processes used to engage people in water issues.

Several authors have noted that when agencies and programs mandate citizen participation, the requirement is often given lip service and a minimum amount of effort. If civic involvement is a requirement, why not spend the time to do it right? One reason is lack of accountability and credit. Gaining successful citizen participation involves hard work. The importance of public sector specialists in their role as facilitators for that work cannot be overstated. It takes time for local groups to organize and start moving. If their efforts are ultimately to be successful and sustainable, agency facilitators need to find interim measures to document accomplishments.

It is appropriate to credit citizens, because a successful local group must ultimately act on its own initiative, but it is also important to recognize the importance and need for agency interventions to bring that about. Even in the most activist community, success absolutely depends on good planning, education, and a commitment to follow through on the part of sponsoring agencies and specialists. A project must justify committing resources to public involvement so managers and staff must receive credit. The watershed project can maintain funding by accountable progress for accomplishments in the social area just as for accomplishments in cost-share funds spent, practices completed, and other traditional watershed interim measures of success.

Prokopy and Floress (Chap. 7) have given us a framework based on categories of participation that can be used to evaluate the success of citizen participation from a project-planning standpoint. These categories also offer public sector projects something to report against as social accomplishments. They give a practical format to, first, describe specifically what type of participation is currently the objective, and then measure the quality and depth of that participation. Their approach makes it possible to specify citizen effect outcomes in a project plan and document accomplishments. It is expected that agency programs will increasingly reward social watershed accomplishment as Prokopy and colleagues have provided a clear framework of measurable social indicators. Their approach will also allow, over time, research to further examine which social outcomes link best to environmental outcomes and that will continue to improve our ability to mobilize the citizen effect.

In small watersheds, it can be much easier to document success in citizen involvement because public agency staff should be closely associated with many of the principal actors. For example, Morton and McGuire's report on the Hewitt Creek performance incentive program found that, in this northeast Iowa watershed, neighbor-to-neighbor interaction brought nearly 70% of watershed farmers into their local project. They have also been able to directly calculate the acreage, including a majority of watershed cropland, that is under performance-based management. As levels of participation go, the Hewitt Creek accomplishments are remarkable. Both attaining these levels and documenting them are made easier by small watershed size.

Targeting

Finally, we propose that successful citizen participation can be evaluated by their contribution to targeting effective management practices in their watersheds. Cost effectiveness of best management practices and how to get the most pollution prevention/remediation for public funds spent on agricultural nonpoint source pollution is an issue of primary concern to agencies nationwide. Surveys and models are used to assess high-priority sites and behaviors to increase program effectiveness. Cost-share levels are adjusted. Public employees are asked to spend additional time with specific landowners. However, the basic human dimension problem is not solved by these analyses and actions. As described in Flora's model, all of these efforts are top-down use of positive or negative force and do nothing to create a climate of cooperation or sustained actions.

Good and bad environmental managers, large and small farms – many or even most operations in any given watershed are likely to have some problem fields or sites from a nonpoint source pollution standpoint. Therefore, when agencies talk about targeting, farmers unavoidably feel that a "target" has been painted on their backs. Those who are conservation minded and proactive will look for ways to avoid agency scrutiny by meeting a minimum standard but they do not become engaged in the big picture issue. Landowners who are "bad actors" are probably not interested in any government program. The missing links in targeting that only citizen participation can supply are public education, support for normative change in the community, neighbor's expectations, and peer pressure. When citizens are educated about pollution and why specific practices are effective, and allowed some decision-making role in managing their watershed, it has been found that they start to practice targeting on their own. Or does it deserve another name? When farmers identify and work on sites in their own operations, or actively recruit each other, they are not targeting – they are setting priorities according to what *they* have learned and values *they* have internalized. Perhaps "re-languaging" the term "targeting" to mean priority setting would more accurately reflect the monitoring and adaptive management processes that targeting represents. Individuals take ownership of their own results, both positive and negative, in the presence of their peers because they have become engaged in working on the problem of an impairment together. A measure of the citizen effect is the extent to which this kind of positive targeting is occurring.

The Watershed Project

Funding for community-based activities is more available than it once was, but the majority of specialists working in watersheds still do not have training for facilitating public engagement. In society at large, we take it for granted that interventions by social scientists ("human dimensions" experts) can make important contributions to community development efforts for economic improvement or quality-of-life issues. Most people, however, still do not consider that the social sciences can make the same contribution to practical management of scientific and environmental problems. Agencies and technical/subject matter specialists in general consistently underestimate what engaged citizens can learn and accomplish in addressing scientific and technical problems.

At some point in every watershed effort, there will be a judgment call by the agencies that control funding about the amount of time and money they will commit to implement the processes that lead to citizen involvement. Involving people is a lot of work. From what we have seen in these chapters, actions that must be taken include some degree of emphasis on *all* of the following:

- Conducting a thorough baseline assessment of citizen ideas, opinions, and practices using surveys (by accepted methods) and other local materials
- Identifying local political and economic power structures to be sure groundwork is laid with the right actors
- Finding key influential people to set the example, share leadership, and champion the cause of water quality
- Conducting, or identifying educators who will conduct, education by participatory discovery that engages citizens with research-based information about impairments and improved management
- Being ready to help organize and interpret (or find those who can) volunteer water monitoring, which is invariably a priority for watershed residents
- Properly and thoroughly planning, organizing, advertising, and conducting local meetings so that residents will actually show up and get engaged
- Being prepared to follow through – being steps ahead at the outset with actions that citizens can sign on to maintain local momentum

- Locating assets to assist with leadership development, to turn interested citizens into committed citizens able to take a substantive role in moving a watershed effort forward
- Assisting watershed groups as they seek funding for local initiatives that are not necessarily in the agency plan or priorities but which support and increase citizen engagement and participation in the those agency priorities

Education and Science Literacy

In many ways, involving citizens in water issues is about creating the citizen-scientist. As we understand better the role of water in the ecosystem and recognize the effects of human activities both large and small, we are better able to choose pathways that protect this valuable resource. Watershed management projects that engage citizens through education and substantive roles in decision making effectively address a central problem of implementing environmental improvement in the USA, which is the need to increase public science literacy. Ownership of monitoring and performance-based management results leads citizens to actively seek science-based knowledge, and to act on it.

In addition to increasing science literacy, watershed projects that involve citizens offer a number of outcomes that benefit not only the project but the community.

- Engaging more people creates local momentum for environmental management and mobilizes both positive and negative community support and peer pressure.
- Solutions can be more effectively targeted because land managers internalize personal environmental priorities. Individuals take action for incremental improvement at their own initiative and recruit their neighbors to do the same. This also means project technical staff make more effective use of their time by focusing on delivery of the expert services they are trained to provide.
- Sustainable actions are produced by replacing "power over" with "power with" thus building increased trust and cooperation among public and private sector individuals and organizations. Power-with relationships make long-term adaptive planning and management possible, strategies that will be needed to implement comprehensive watershed plans as envisioned, for example, by the total maximum daily load (TMDL) process.
- Engaged citizens are more likely to take advantage of public cost-share programs and recruit their neighbors to do so, thereby making more effective use of public funds for environmental improvement.
- Additional interim metrics are available for documenting watershed improvement success. Projects that seek to involve citizens can report against numerous social indicators as well as environmental indicators.
- The way for all types of environmental initiatives is smoothed, even those that begin with regulatory programs external to the community, because of increased trust and communication among citizens and the public sector.

Why Citizen Involvement?

Why should we seek citizen involvement and leadership in watersheds? Because it works – is the short answer to this ultimate question. As the studies and examples presented in this volume show, citizen participation leads to better outcomes in many of the interim measures that public sector projects need to document success and continue their programs. This situation makes the effort to involve citizens a win–win situation for both public agencies and the community. Cultivating local champions, developing citizen leadership, and providing flexible funding can also lead to better outcomes in ways that public agencies never hoped to impact, outcomes that lead to a fundamental change in the culture of agriculture and rural landusers. These outcomes stem from individuals and groups internalizing environmental concerns and adopting personal environmental goals, in addition to economic and social goals, in their land management. When offered opportunities to learn and engage each other, citizens can and do reconstruct their beliefs, change their behaviors, and alter their local culture to be one of environmental stewardship.

The research findings and practice experiences presented here are only a fraction of what we yet need to know to effectively leverage the citizen effect in addressing natural resource issues. Human action is a key variable in the restoration, protection, and improvement of water resources. Future research on involving citizens in watershed management must engage the whole social science continuum from structural theories of society to social-psychology. We must understand how individuals and groups nested in communities of interest and communities of place are linked to their natural environment. This means applying theories of group dynamics, collective action, social movements, social structure, social organization, human agency, social relationships, and individual and collective decision making to the complex intersections of human society and nature. There is much yet to be learned about the human and social dimension of getting to better water outcomes. However, there is solid preliminary evidence that citizen involvement makes a nonpoint source watershed investment more effective in the long run because local interest, commitment, and peer pressure can keep planning and implementation in motion even when specific funding source(s) go away.

Our society has agreed that protection and improvement of water resources for drinking water, economic, recreational, and aesthetic purposes is a legitimate use of public expenditures. The scientific community knows that, regarding water quality and availability, priority environmental problems and technical approaches will evolve and change. The regulatory community knows that programs of positive and negative sanctions will evolve and change. Local residents know that social and economic influences on their communities will evolve and change. Among these changing forces, citizen engagement and leadership on watershed management issues – the citizen effect – can and should be cultivated as a unifying energy that makes moving toward better watershed outcomes an adaptive and sustainable process for our shared future.

Bibliography

Ad Hoc Task Force on Agriculture and New York City Watershed Regulations (1991) Policy group of recommendations, December 1991 (unpublished manuscript)

Ajzen I (2001) Nature and operation of attitudes. Annu Rev Psychol 52:27–58

Alexander RB, Smith RA, Schwarz GE, Boyer EW, Nolan JV, Brakebill JW (2008) Differences in phosphorus and nitrogen delivery to the Gulf of Mexico from the Mississippi River Basin. Environ Sci Technol 42(3):822–830; supplemental material pages S22–S25. http://pubs.acs.org/doi/suppl/10.1021/es0716103, verified 11-14-2010

Alinsky S (1972) Rules for radical. Random House, New York, NY

Allen BL, Morton LW (2006) Shared leadership practices among non-profits in Iowa. J Ext 44(6):1–12. http://www.joe.org, verified 11-14-2010

Almond GA, Verba S (1989) The civic culture revisited. Sage, London, UK

Anderson TL, McChesney FS (2003) The economic approach to property rights. In: Anderson TL, McChesney FS (eds) Property rights: cooperation, conflict and law. Princeton University Press, Princeton, NJ, pp 1–11

Arcury TA, Christianson EH (1990) Environmental worldview in response to environmental problems. Environ Behav 22(3):387–407

Aristotle (1996) The politics and the constitution of Athens. Cambridge University Press, Cambridge, UK

Armitage DR (2005) Community-based narwhal management in Nunavut, Canada: change, uncertainty and adaptation. Soc Nat Resour 18:715–731

Arnstein SR (1969) A ladder of citizen participation. J Am Inst Plann 35(4):216–224

Bakhtin MM (1981) The dialogic imagination: four essays. University of Texas Press, Austin, TX

Ball C, Knight B (1999) Why we must listen to citizens. In: K. Naidoo (ed) Civil society at the millennium, chapter 2. Kumarian, West Hartford, CT

Bandura A (1971) Social learning theory. General Learning, Morristown, NJ

Barber BR (1996) Jihad vs. McWorld. Ballantine Books, New York, NY

Barden CJ (2003) Lessons learned from collaboration with the Potawatomi Nation. Presented at 6th International Union of Forestry Research Organizations (IUFRO) extension working party symposium, Troutdale, OR, 30 September 2003

Barham E (2001) Ecological boundaries as community boundaries: the politics of watersheds. Soc Nat Resour 14:181–191

Basso E (1990) Introduction: discourse as an integrating concept in anthropology and folklore research. In: Basso E (ed) Native Latin American cultures through their discourse. Folklore Institute, Indiana University, Bloomington, IN, pp 3–10

Beeman P (2004) Panel votes to let limited on river pollution stand. The Des Moines Register, p 3B, 16 March 2004

Beierle TC, Cayford J (2002) Democracy in practice: public participation in environmental decisions. Resources for the Future, Washington, DC

L.W. Morton and S.S. Brown (eds.), *Pathways for Getting to Better Water Quality: The Citizen Effect*, DOI 10.1007/978-1-4419-7282-8,
© Springer Science+Business Media, LLC 2011

Bidwell RD, Ryan CM (2006) Collaborative partnership design: the implications of organizational affiliation for watershed partnerships. Soc Nat Res 19:827–843

Blackard K, Gibson JW (2002) Capitalizing on conflict: strategies and practices of turning conflict to synergy in organizations. Davies-Black, Palo Alto, CA

Blomquist W, Schlager E (2005) Political pitfalls of integrated watershed management. Soc Nat Resour 18:101–117

Bonnell JE, Koontz TM (2007) Stumbling forward: the organizational challenges of building and sustaining collaborative watershed management. Soc Nat Resour 20:153–167

Bourdieu P (1986) The forms of capital. In: Richardson JD (ed) Handbook of theory and research for the sociology of education. Greenwood, Westport, CT, pp 241–258

Born SM, Genskow KD (2000) The watershed approach: an empirical assessment of innovation in environmental management. National Academy of Public Administration Research Paper Number 7. http://www.napawash.org, verified 11-14-2010

Boyte HC (2004) Everyday politics: reconnecting citizens and public life. University of Pennsylvania Press, Philadelphia, PA

Boyte HC (2008) The citizen solution: how you can make a difference. Minnesota Historical Society, St. Paul, MN

Brehm JM, Eisenhauer BW, Krannich RS (2006) Community attachments as predictors of local environmental concern: the case for multiple dimensions of attachment. Am Behav Sci 50(2):142–165

Brick P, Snow D, Van De Wetering S (eds) (2001) Across the great divide: explorations in collaborative conservation and the American West. Island Press, Washington, DC

Bruins J (1999) Social power and influence tactics: a theoretical introduction. J Soc Issues 55(1):7–14

Bureau of Reclamation (2008). http://www.usbr.gov/dataweb/html/Wichita.html#general, accessed 10-1-2009

Busch L, Lacy WB, Burkhardt J, Hemken D, Moraga-Rojel J, Koponen T, de Souza Silva J (1995) Making nature shaping culture. University of Nebraska Press, Lincoln, NE

Buttel FH, Gillespie GW Jr, Larson OW III, Harris CK (1981) The social bases of agrarian environmentalism: a comparative analysis of New York and Michigan farm operators. Rural Sociol 46(3):391–410

Buttel FH, Flinn WL (1976) Economic growth versus the environment: survey evidence. Soc Sci Q 57:410–420

Buttel FH, Humphrey CR (2002) Sociological theory and the natural environment. In: Dunlap RE, Michelson W (eds) Handbook of environmental sociology, chapter 2. Greenwood, Westport, CT

Cable S, Degutis B (1997) Movement outcomes and dimensions of social change: the multiple effects of local mobilizations. Curr Sociol 45(3):121–135

Cawley RM (1993) Federal land, western anger: the sagebrush rebellion and environmental politics. University Press of Kansas, Lawrence, KS

Chambers R (1983) Rural development: putting the last first. Longman, London, UK

Christensen LA, Norris PE (1983) Soil conservation and water quality improvement: what farmers think. J Soil Water Conserv 38:15–20

Christenson JA, Robinson JW (1989) Community development in perspective. Iowa State University Press, Ames, IA

City of Wichita (2009) Cheney reservoir environmental assessment summary. http://www.wichita.gov/CityOffices/WaterAndSewer/ProductionAndPumping/Cheney.htm, accessed, 10-3-2009

Cochnar J, Perkins B (2005) Service proposes protection of the Salt Creek Tiger Beetle. News Release: U.S. Fish and Wildlife Service, 22 March 2007. http://mountain-prairie.fws.gov/pressrel/05-06.htm, verified 11-14-2010

Cohen JL (1999) American civil society talk. In: Fullinwider RK (ed) Civil society, democracy, and civic renewal, chapter 3. Rowman and Littlefield, New York

Conca K (2006) Governing water: contentious transnational politics and global institution building. MIT, Cambridge, MA

Cook KS, Cheshire C, Gerbasi A (2006) Power, dependence, and social exchange. In: Burke PJ (ed) Contemporary social psychological theories, chapter 9. Stanford University Press, Stanford, CA

Coplin WD, O'Leary MK (1972) Everyman's prince: a guide to understanding your political problems. Duxbury, North Scituate, MA

Correll SJ, Ridgeway CL (2006) Expectation states theory. In: Delamater J (ed) Handbook of social psychology, chapter 2. Springer Science+Business Media, LLC, New York, NY

Coughenour CM (2003) Innovating conservation agriculture: the case of no-till cropping. Rural Sociol 68(2):278–304

Cranford EE, Kleinschmit J (2007) Fishbowls in the field: using listening to join farmers, ranchers, and educations in advancing sustainable agriculture. CARI: Center for Applied Rural Innovation at the University of Nebraska, Lincoln. http://digitalcommons.unl.edu/caripubs/39, verified 11-14-2010

Cunningham D (1985) Villains, miscreants, and the salt of the earth. NEBRASKAland Mag July, pp 14–19, 45–47

Daniels AC, Daniels JE (2007) Measure of a leader. McGraw-Hill, New York, NY

de Geus AP (1994) Modeling to predict or to learn? In: Morecroft JDW, Sterman JD (eds) Foreword in Modeling for Learning Organizations. Productivity Press, Portland, OR

Deloria V Jr, Wildcat D (2001) Power and place: Indian education in America. Fulcrum Resources, Golden, CO

DeMey M (1982) The cognitive paradigm. D. Reidel, Dordrecht, Holland

Dennison WC, Lookingbill TR, Carruthers TJB, Hawkey JM, Carter SL (2007) An eye-opening approach to developing and communicating integrated environmental assessments. Front Ecol Environ 5(6):307–314

Devlin D, Nelson N, French L, Miller H, Barnes P, Frees L (2008) Conservation practice implementation history and trends. Kansas State University Agricultural Experiment Station and Cooperative Extension Service, Publication # EP157, Manhattan, KS. http://www.oznet.ksu.edu/library/h20ql2/ep157.pdf

Dietz T, Stern PC (eds) (2008) Public participation in environmental assessment and decision making, Committee on the Human Dimensions of Global Change, Division of Behavioral and Social Sciences and Education. National Research Council of the National Academies. The National Academies Press, Washington, DC. http://www.nap.edu

Dillman DA (2000) Mail and internet surveys, 2nd edn. Wiley, New York, NY

Dillman DA, Christenson JA (1972) The public value for pollution control. In: Burch WR Jr, Creek NH Jr, Taylor L (eds) Social behavior, natural resources and the environment. Harper and Row, New York, NY, pp 237–256

Downing B (2004) Farmers help Sugar Creek grow cleaner. Akron Beacon J September 24, B1

Drewal MT (1992) Yoruba ritual: performers, play, agency. Indiana University Press, Bloomington, IN

Ducey J (1985) Nebraska's salt basin: going, going, nearly gone. NEBRASKAland Mag July, pp 20–25

Dunlap RE, Jones RE (2002) Environmental concern: conceptual and measurement issues. In: Dunlap RE, Michelson W (eds) Handbook of environmental sociology. Greenwood, Westport, CT

Dunlap RE, Van Liere KD (1978) The 'new environmental paradigm'. J Environ Educ 9:10–19

Dunlap RE, Van Liere K, Mertig AG, Jones RE (2000) New trends in measuring environmental attitudes: measuring endorsement of the new ecological paradigm: a revised NEP scale. J Soc Issues 56(3):425–442

Dunlap R (1998) Lay perceptions of global risk: public views of global warming in cross-national context. Int Sociol 13(4):473–498

Ellen RF (1984) Ethnographic research: a guide to general conduct. Academic, London, UK

Epstein PD, Coates PM, Wray LD, Swain D (2006) Results that matter: improving communities by engaging citizens, measuring performance and getting things done. Jossey-Bass, San Francisco, CA

Farrar J (2005) Preserving the last of the least: a partnership of organizations is working to protect the Saline Wetlands. NEBRASKAland Mag January–February, p 46

Ferreyra C, de Loe RC, Kreutzwiser RD (2008) Imagined communities, contested watersheds: challenges to integrated water resources management in agricultural areas. J Rural Stud 24:304–321

Filipovitch AJ (2005) PRINCE analysis. http://Krypton.mnsu.edu/~tony/courses/609/Frame/PRINCE.html. Retrieved 6 Apr 2008

Finnegan MC (1997) New York City's watershed agreement: a lesson in sharing responsibility. Pace Environ Law Rev 14:577–644

Fischer F (2000) Citizens, experts, and the environment: the politics of local knowledge. Duke University Press, Durham, NC

Fischer F (2005) Citizens, experts, and the environment: the politics of local knowledge. Duke University Press, Durham, NC

Flora CB (2004) Social aspects of small water systems. J Contemp Water Res Educ 128:6–12

Folke C, Hahn T, Olsson P, Norberg J (2005) Adaptive governance of social-ecological systems. Annu Rev Environ Resour 30:441–473

Fox W (1995) Toward a transpersonal ecology: developing new foundations for environmentalism. State University of New York Press, New York, NY

French JRP Jr, Raven B (1968) The bases of social power. In: Cartwright D, Zander A (eds) Group dynamics: research and theory. Harper & Row, New York, NY, pp 259–269

Freudenburg WR (1991) Rural-urban differences in environmental concern: a closer look. Sociol Inq 61(2):167–198

Gamson W (1975) The strategy of social protest. Dorsey, Homewood, IL

Gattin TD (1995) Cheney Reservoir: creating and preserving a surface water supply for the City of Wichita. Unpublished Masters' Thesis. Hugo Wall Center for Urban Studies, Wichita State University, Wichita, KS

Genskow, Ken and Linda Prokopy (eds.). 2008. *The Social Indicator Planning and Evaluation System (SIPES) for Nonpoint Source Management: A Handbook for Projects in USEPA Region 5.* Great Lakes Regional Water Program. Publication Number: GLRWP-08-SI01

Goffman E (1974) Frame analysis: an essay on the organization of experience. Harvard University Press, Cambridge, MA

Goode E, Ben-Yehuda N (1994) Moral panics: the social construction of deviance. Blackwell, Oxford, UK

Graham LR (2000) The one who created the sea: tellings, meanings and inter-textuality in the translation of xavante narrative. In: Sammons K, Sherzer J (eds) Translating Native American verbal art: ethnopoetics and ethnography of speaking. Smithsonian Institution, Washington, DC, pp 252–271

Greenberg MR (2005) Concern about environmental pollution: how much difference do race and ethnicity make? A New Jersey case study. Environ Health Perspect 113:369–374

Grumbine RE (1997) Reflections on 'what is ecosystem management?'. Conserv Biol 11(1):41–47

Habron G (2003) Role of adaptive management of watershed councils. Environ Manag 31(1):29–41

Hand CM, Van Liere K (1984) Religion, mastery-over-nature, and environmental concern. Soc Forces 2:555–570

USEPA (2005) Handbook for developing watershed plans to restore and protection our waters. http://www.epa.gov/owow/nps/cwact.html, accessed 10-3-2009

Hayes B (2001) Gender, scientific knowledge and attitudes toward the environment: a cross-national analysis. Polit Res Q 54:657–671

Hechter M (1987) Principles of group solidarity. University of California Press, Berkeley, CA

Heinz Center (2008) The state of the nation's ecosystems 2008: measuring the lands, waters, and living resources of the United States. The H. John Heinz III Center for Science, Economy, and the Environment. Island Press, Washington, DC

Higdon F, Brasier K, Stedman R, Lee B, Sherman S (2005) Assessment of community watershed organizations in rural Pennsylvania. Center for Rural Pennsylvania. http://www.ruralpa.org/watersheds_higdon.pdf, verified 11-14-2010

Hobbes T (1651/1962). In: M Oakshott (ed) Leviathan. Collier-Macmillan, London, UK

Hufford DB (1982) The ayes have it, Wichita Water Department, a history 1882–1982. Frank Wright-Josten, Wichita, KS

Hunter F (1953/1974) Community power structure. In: Hawley WD, Wirt FM (eds) The search for community power. Prentice-Hall, Englewood Cliffs, NJ, pp 252–271

Innes JE (1994) Knowledge and public policy: the search for meaningful indicators, 2nd edn. Transaction, London, UK

Iowa Department of Natural Resources (2002). http://programs.iowadnr.gov/adbnet/segment. aspx?sid=110, verified 11-14-2010

Irvin RA, Stansbury J (2004) Citizen participation in decision making: is it worth the effort? Public Adm Rev 64(1):55–65

Irvine JT (1996) Shadow conversations: the indeterminacy of participant roles. In: Silverstein M, Urban G (eds) Natural histories of discourse. University of Chicago Press, Chicago, IL, pp 131–159

Jackson-Smith D, Kreuter U, Krannich R (2005) Understanding the multidimensionality of property rights orientations: evidence from Utah and Texas ranchers. Soc Nat Resour 18:587–610

Janoski T (1998) Citizenship and civil society. Cambridge University Press, Cambridge, UK

Janov J (1994) The inventive organization: hope and daring at work. Jossey-Bass, San Francisco, CA

Johnson CY, Bowker JM, Cordell HK (2004) Ethnic variation in environmental belief and behavior: an examination of the new ecological paradigm in a social psychological context. Environ Behav 36(2):157–186

Jones RE, Dunlap RE (1992) The social bases of environmental concern: have they changed over time? Rural Sociol 57:28–47

Kaner S (2007) Facilitator's guide to participatory decision-making, 2nd edn. Jossey Bass, San Francisco, CA

Kellert SR, Mehta JN, Ebbin SA, Lichtenfeld LL (2000) Community natural resource management: promise, rhetoric, and reality. Soc Nat Resour 13:705–715

Kenny JF, Hansen CV (2004) Water use in Kansas, 1990–2000. U.S. Geological Survey. http:// pubs.water.usgs.gov/fs20043133, verified 11-14-2010

Knox TM (1821/1967) Hegel's philosophy of right (Transl.). Oxford University Press, Oxford, UK

Koontz TM, Steelman TA, Carmin J, Korfmacher KS, Moseley C, Thomas CW (2004) Collaborative environmental management: what roles for government? Resources for the Future, Washington, DC

LaGrange T, Genrich T, Johnson G, Schulz D (2002) Implementation plan for the conservation of Nebraska's Eastern Saline Wetlands (draft). Eastern Saline Wetlands Project

Leach WD, Pelkey NW (2001) Making watershed partnerships work: a review of the empirical literature. J Water Resour Plann Manag 127(6):378–385

Leach WD, Pelkey NW, Sabatier PA (2002) Stakeholder partnerships as collaborative policymaking: evaluation criteria applied to watershed management in California and Washington. J Policy Anal Manag 21(4):645–670

Leas SB (1982) Leadership and conflict. Abingdon, Nashville, TN

Lee B (2005) Pennsylvania community watershed organizations: form and function. Doctoral dissertation, The Pennsylvania State University

Leftwich A (1984) What is politics? Basil Blackwell, Oxford, UK

Leopold A (1949) A sand county almanac and sketches here and there. Oxford University Press, New York, NY

Leopold LB (ed) (1993) In round river; from the journals of Aldo Leopold. Oxford University Press, New York, NY, pp 156–157

Lewin K (1951) Field theory in the social sciences. Harper and Row, New York, NY

Lockeretz W (1990) What have we learned about who conserves soil? J Soil Water Conserv 445(1):132–136

Lowe GD, Pinhey TK (1982) Rural-urban differences in support for environmental protection. Rural Sociol 47:114–128

Lubell M (2004) Collaborative watershed management: a view from the grassroots. Policy Stud J 32:341–361

Lundmark C (2007) The new ecological paradigm revisited: anchoring the NEP scale in environmental ethics. Environ Educ Res 13:329–347

Machiavelli N (1532) A Florentine, Italy public servant and political theorist who wrote Principe (The Prince) a political treatise

Mahan RA, Bollman SR (1968) Education or information giving? J Coop Ext Summer:100–106

Mannheim B, Tedlock D (1995) Introduction. In: Tedlock D, Mannheim B (eds) The dialogic emergence of culture. University of Illinois Press, Urbana, IL, pp 1–32

Maquoketa Quarterly Reports (1999) EPA Region VII water quality cooperative agreement. Iowa State University, Ames, IA (October and December)

Mathews F (1994) The ecological self. Routledge, London, UK

Maxwell JC (1993) Developing the leader within you. Thomas Nelson, Nashville, TN

McCarthy J (2005) Devolution in the woods: community forestry as hybrid neoliberalism. Environ Plann A 37:995–1014

McCool SF, Guthrie K (2001) Mapping the dimensions of successful public participation in messy natural resources management situations. Soc Nat Resour 14:309–323

McCown RL (2002) Changing systems of supporting farmers' decisions: problems, paradigms, and prospects. Agric Syst 74:179–220

McCown RL (2005) New thinking about farmer decision makers. In: Hatfield JL (ed) The farmer's decision. Soil and Water Conservation Society, Ankeny, IA, pp 11–44

McNicoll G (1995) Demography in the unmaking of civil society. The Population Council. Working Paper No. 79. Canberra: Australian National University

Meyer JW, Jepperson RL (2000) The 'actors' of modern society: the cultural construction of social agency. Sociol Theory 18(1):100–120

Milfont TL, Duckitt J, Cameron LD (2006) A cross-cultural study of environmental motive concerns and their implications for proenvironmental behavior. Environ Behav 38(6):745–767

Miller D (1987) The Blackwell encyclopedia of political thought. Basil Blackwell, Oxford, UK

Minkler M, Wallerstein N (1999) Improving health through community organization and community building. In: Minkler M (ed) Community organizing and community building for health, chapter 3. Rutgers University Press, New Brunswick, NJ

Minkler M (1999) Community organizing and community building for health. Rutgers University Press, New Brunswick, NJ

Mitchell RC (1991) From conservation to environmental movement: the development of the modern environmental lobbies. In: Lacey MJ (ed) Government and environmental politics. Johns Hopkins University Press, Baltimore, MD, pp 81–114

Mohai P, Bryant B (1998) Is there a 'race' effect on concern for environmental quality? Public Opin Q 62:475–505

Moore EA, Koontz TM (2003) A typology of collaborative watershed groups: citizen-based, agency-based, and mixed partnerships. Soc Nat Resour 16(5):451–460

Moore R, Parker J, Weaver M (2008) Agricultural sustainability, water pollution, and governmental regulations: lessons from the Sugar Creek farmers in Ohio. Cult Agric 30(1–2):3–16

Moote A, Lowe K (2008) What to expect from collaboration in natural resource management: a research synthesis for practitioners. Ecological Restoration Institute, Flagstaff, AZ

Morecroft JDW (1994) Executive knowledge, models, and learning. In: Morecroft JDW, Sterman JD (eds) Modeling for learning organizations, chapter 1. Productivity Press, Portland, OR

Morton LW (2003a) Civic watershed communities: walking toward justice: democratization in rural life. Res Rural Sociol Dev 9:121–134

Morton LW (2003b) Small town services and facilities: the influence of social capital and civic structure on perceptions of quality. City Community 2(2):99–118

Morton LW (2003c) Civic structure. In: Christensen K, Levinson D (eds) Encyclopedia of community: from the village to the virtual world. Sage, Thousand Oaks, CA, pp 179–182

Morton LW (2003c) Civic watershed communities. In: Bell MM, Hendricks FT, Bacal A (eds) Walking towards justice: democratization in rural life, chapter 8. Research in rural sociology and development, vol 9. JAI/Elsevier, Amsterdam, The Netherlands, pp 121–134

Morton LW (2008) The role of civic structure in achieving performance based watershed management. Soc Nat Res 21(9):751–766

Morton LW, Brown S (2007) Water issues in the four state heartland Region: a survey of public perceptions and attitudes about water. The Heartland Regional Water Coordination Initiative Bulletin #SP289 Iowa State University Extension. http://www.heartlandwq.iastate.edu

Morton LW, Miller G, Rodecap J, Brown S (2006) Performance-based environmental management: The Hewitt Creek Model. Iowa State University Extension, Publication PM 2013, Ames, IA. http://www.extension.iastate.edu/Publications/PM2013.pdf

Morton LW, Padgitt S (2005) Selecting socio-economic metrics for watershed management. Environ Monit Assess 103:83–98

Morton LW, Weng CY (2009) Getting to better water quality outcomes: the promise & challenge of the citizen effect. Agric Human Values 26(1):83–94

Myerhoff BG (1992) Life history among the elderly: performance, visibility, and remembering. In: Kaminsky M (ed) Remembered lives: the work of ritual, storytelling, and growing older. University of Michigan Press, Ann Arbor, MI, pp 231–340

Napier TL, Tucker M, McCarter S (2000) Adoption of conservation production systems in three midwest watersheds. J Soil Water Conserv 55(2):123–134

Nassauer JI (1989) Agricultural policy and aesthetic objectives. J Soil Water Conserv 44(5):384–387

National Research Council (NRC) (2000) Watershed management for potable water supply: assessing the New York City strategy. National Academy Press, Washington, DC

National Research Council (2008) Public participation in environmental assessment and decision-making. In: Dietz T, Stern PC, Committee on the Human Dimensions of Global Change, Division of Behavioral and Social Sciences and Education (ed) Panel on public participation in environmental assessment and decision-making. The National Academies Press, Washington, DC

Nelson MW (2008) Wichita Project – Cheney Division, Kansas. Bureau of Reclamation. http://www.usbr.gov/dataweb/html/wichita.html#general, accessed 8-16-2008

Nelson N, Mankin K, Langemeier M, Devlin D, Barnes P, Selfa T, Hargrove W (2006) Assessing the impact of a strategic approach to implementation of conservation practices. USDA-CSREES Conservation Effects Assessment Program, 2006–2009

New York City Department of Environmental Protection (DEP) (1990) Discussion draft of proposed regulations for the protection from contamination, degradation and pollution of the New York City water supply and its sources. NYCDEP, Corona, NY

New York City Department of Environmental Protection (DEP) (1994) Draft generic environmental impact statement for the draft watershed regulations for the protection from contamination, degradation, and pollution of the New York City water supply and its sources. NYCDEP, New York, NY

New York City Watershed Memorandum of Agreement (MOA). January 1997 (unpublished manuscript)

Nolon JR (1993) The erosion of home rule through the emergence of state interests in land use control. Pace Environ Law Rev 10(2):497–562

Nooney JG, Woodrum E, Hoban TJ, Clifford WB (2003) Environmental worldview and behavior: consequences of dimensionality in a survey of North Carolinians. Environ Behav 35(6):763–783

Novotny V, Chesters G (1981) Handbook of nonpoint pollution. Van Nostrand Reinhold, New York, NY

Nowak P, Korsching PF (1998) The human dimension of soil and water conservation: a historical and methodological perspective. In: Pierce FJ, Frye WW (eds) Advances in soil and water conservation. Ann Arbor Press, Chelsea, MI, pp 231–340

O'Neill K (2005) Can watershed management unite town and country? Soc Nat Resour 18(3):241–253

Ohio Environmental Protection Agency (2000) Biological and water quality study of Sugar Creek, 1998. OEPA Technical Report MAS/1999-12-4

Ohio Environmental Protection Agency (2002) Total maximum daily loads for the Sugar Creek Basin. http://www.epa.state.oh.us/dsw/tmdl/SugarCreek.html, accessed 9-7-2008

Olofsson A, Ohman S (2006) General beliefs and environmental concern: transatlantic comparisons. Environ Behav 38:768–790

Pagdee A, Kim YS, Daugherty PJ (2006) What makes community forestry management successful: a meta-study from community forests throughout the world. Soc Nat Resour 19:33–52

Parisi D, Taquino M, Grice SM, Gill DA (2004) Civic responsibility and the environment: linking local conditions to community environmental activeness. Soc Nat Resour 17(2): 97–112

Parker J, Moore R, Weaver M (2007) Land tenure as a variable in community based watershed projects: some lessons from the Sugar Creek Watershed, Wayne and Holmes Counties, Ohio. Soc Nat Resour 20(9):815–833

Pfeffer J (1994) Managing with power: politics and influence in organizations. Harvard Business Press, Boston, MA

Pfeffer MJ (2003) The watershed as community. In: Christensen K, Levinson D (eds) Encyclopedia of community: from the village to the virtual world. Sage, Thousand Oaks, CA

Pfeffer MJ, Wagenet LP (1999) Planning for environmental responsibility and equity: a critical appraisal of rural/urban relations in the New York City Watershed. In: Lapping MB, Furuseth O (eds) Contested countryside: the rural urban fringe of North America. Ashgate, Brookfield, VT, pp 179–206

Pfeffer MJ, Stycos JM, Glenna L, Atobelli J (2001a) Forging new connections between agriculture and the city. In: Solbrig OT, Paarlberg R, di Castri F (eds) Globalization and the rural environment. Harvard University Press, Cambridge, MA, pp 419–446

Pfeffer MJ, Schelhas JW, Day LA (2001b) Forest conservation, value conflict, and interest formation in a Honduran National Park. Rural Sociol 66(3):382–402

Pfeffer MJ, Wagenet LP, Sydenstricker-Neto J, Meola C (2005) Reconciling different land use value spheres: an example at the rural/urban interface. In: Goetz S, Shortle J, Bergstrom J (eds) Land use problems and conflicts: causes, consequences and solutions. Routledge, London, UK, pp 186–201

Plato (1987) The republic. Penguin, London, UK

Platt RH, Barten PK, Pfeffer MJ (2000) A full, clean glass? Managing New York City's watersheds. Environment 42(5):9–20

Polanyi M (1958) Personal knowledge: towards a post-critical philosophy. University of Chicago Press, Chicago, IL

Portes A (1998) Social capital: its origins and applications in modern sociology. Annu Rev Sociol 24:1–12

Prokopy L, Floress K, Klotthor-Weinkauf D, Baumgart-Gertz A (2008) Determinants of agricultural best management practice adoption: evidence from the literature. J Soil Water Conserv 63(5):300–311

Putnam RD (1993) Making democracy work. Princeton University Press, Princeton, NJ

Putnam RD (2000) Bowling alone: the collapse and revival of American community. Simon & Schuster, New York, NY

Ravnborg HM, del Pilar Guerrero M (1999) Collective action in watershed management-experiences from the Andean Hillsides. Agric Human Values 16:257–266

Rhoads BL, Wilson D, Urban M, Herricks EE (1999) Interaction between scientists and nonscientists in community-based watershed management: emergence of the concept of stream naturalization. Environ Manag 24(3):297–308

Ridgeway CL, Correll SJ (2006) Consensus and the creation of status beliefs. Soc Forces 85(1):431–453

Rodecap J (2008) Hewitt Creek Iowa Watershed Review Board progress report. Progress report to the Iowa Watershed Improvement Review Board, Iowa Department of Agriculture and Land Stewardship, Des Moines, IA

Rogers EM (1962, 1971, 1983, 1995, 2003) Diffusion of innovations. Free Press, New York, NY

Routley R (1973) Is there a need for a new, an environmental, ethic? Proceedings of the world congress of philosophy, Varna, Bulgaria, pp 205–210

Ryan RL, Erickson DL, deYoung R (2003) Farmers' motivations for adopting conservation practices along riparian zones in a midwestern agricultural watershed. J Environ Plann Manag 46(1):19–37

Sabatier PA, Jenkins-Smith HC (eds) (1993) Policy change and policy learning: an advocacy coalition approach. Westview, Boulder, CO

Sabatier PA, Jenkins-Smith HC (1999) The advocacy coalition framework: an assessment. In: Sabatier PA (ed) Theories of the policy process. Westview, Boulder, CO, pp 117–166

Sabatier PA, Focht W, Lubell M, Trachtenberg Z, Vedlitz A, Matlock M (eds) (2005) Swimming upstream: collaborative approaches to watershed management. MIT, Cambridge, MA

Satria A, Matsuda Y, Sano M (2006) Questioning community based coral reef management systems: case study of Awig-Awig in Gili Indah, Indonesia. Environ Dev Sustain 8:99–118

Schilling KE, Wolter CF (2001) Contribution of base flow to nonpoint source pollution loads in an agricultural watershed. Groundwater 39(1):49–58

Schnaiberg A (2001) Environmental movements since love canal: hope, despair and [im]mobilization? Buffalo Environmental Law Journal Spring 8:256–269

Schneeweiss J (1997) Watershed protection strategies: a case study of the New York City Watershed in light of the 1996 amendments to the Safe Drinking Water Act. Villanova Environ Law J 9:77–119

Secchi S, Babcock BA (2007) Impact of high crop prices on environmental quality: a case of Iowa and the Conservation Reserve Program. Working paper 07-WP447 Center for Agricultural and Rural Development, Ames, IA. http://www.card.iastate.edu

Shepard R (1999) Making our nonpoint source pollution education programs effective. J Ext 37(5). http://www.joe.org/joe/1999october/a2.html, accessed 2-23-2009

Simon H (1979) From the substantive to procedural rationality. In: Han F, Hollis M (eds) Philosophy and economy theory. Oxford University Press, New York, NY, pp 65–68

Sirianni C, Friedland LA (2005) The civic renewal movement: community-building and democracy in the United States. Charles F. Kettering Foundation, Dayton, OH

Skocpol T, Fiorina MP (1999) Civic engagement in American democracy. Brookings Institution Press, Washington, DC

Spaling H (2003) Innovation in environmental assessment of community-based projects in sub-Saharan Africa. Can Geogr 47(2):151–168

Stave KA (1998) Water, land, and people: the social ecology of conflict over New York City's watershed protection efforts in the catskill mountain region. PhD Dissertation, Yale University, New York

Stedman R, Lee B, Brasier K, Weigle J, Higdon F (2009) Cleaning up water? Or building rural community? Community watershed organizations in Pennsylvania. Rural Sociol 74(2):178–200

Steins NA, Edwards VM (1999) Synthesis: platforms for collective action in multiple-use common-pool resources. Agric Human Values 16:309–315

Stern PC, Young OR, Druckman D (1992) Global environmental change: understanding the human dimensions. National Academy Press, Washington, DC

Stone D (2001) Policy paradox: the art of political decision making. WW Norton, New York, NY

Strange CJ (2007) Facing the brink without crossing it. Bioscience 57(11):920–926

Sullivan WM (1999) Making civil society work: democracy as a problem of civic cooperation. In: Fullinwider RK (ed) Civil society, democracy, and civic renewal, chapter 2. Rowman and Littlefield, New York, NY

Switzer JV (1997) Green backlash: the history and politics of environmental opposition in the U.S. Lynne Rienner, Boulder, CO

Taylor DE (2005) American environmentalism: the role of race, class and gender in shaping activism 1820–1995. In: King L, McCarthy D (eds) Environmental sociology: from analysis to action. Rowman & Littlefield, Lanham, MD, pp 87–106

Tremblay KR, Dunlap RE (1978) Rural urban residence and concern with environmental quality: a replication and extension. Rural Sociol 43:474–491

U.S. Department of Agriculture National Agricultural Statistics Service (2007) Census of agriculture. Washington, DC. http://www.agcensus.usda.gov/Publications/2007, verified 11-14-2010

U.S. EPA (2008) Handbook for developing watershed plans to restore and protect our waters. USEPA. Office of Water Nonpoint Source Control Branch. EPA 841-B-050005

U.S. Fish and Wildlife Service (2005) Salt Creek Tiger Beetle, 22 March 2007. http://mountain-prairie.fws.gov/species/invertebrates/saltcreektiger/index.htm, verified 11-14-2010

U.S. Geological Survey (2009). http://www.usgs.gov/newsroom/article.asp?ID=2240&from=rss_home, verified 11-14-2010

van Es JC, Notier P (1988) No-till farming in the United States: research and policy environment in the development and utilization of an innovation. Soc Nat Resour 1:93–110

Van Liere KD, Dunlap RE (1980) The social bases of environmental concern: a review of hypotheses, explanations, and empirical evidence. Public Opin Q 44:181–197

Voloshinov VN, Matejka L, Titunik IR (1986) Marxism and the philosophy of language. Harvard University Press, Cambridge, MA

Vorkinn M, Riese H (2001) Environmental concern in a local context: the significance of place attachment. Environ Behav 33(2):249–263

Wagenet L, Pfeffer MJ (2007) Organizing citizen engagement for democratic environmental planning. Soc Nat Resour 20(9):801–813

Warren ME (2001) Democracy and association. Princeton University Press, Princeton, NJ

Weaver M, Moore R (2004) Generating and sustaining collaborative decision-making in watershed groups. Presented at 67th Annual Meeting of the Rural Sociological Society, Sacramento, 11–15 August 2004

Weaver M, Moore R, Parker J (2005) Understanding grassroots stakeholders and grassroots stakeholder groups: the view from the grassroots in the Upper Sugar Creek. Presented at the 2005 Annual Meeting of the American Political Science Association, Washington, DC, 1–4 September 2005. http://convention2.allacademic.com/one/apsa/apsa05/, accessed 8-28-2008

Weber EP (2000) A new vanguard for the environment: grass-roots ecosystem management as a new environmental movement. Soc Nat Resour 13:237–259

Weber EP (2003) Bringing society back in: grassroots ecosystem management, accountability, and sustainable communities. MIT, Cambridge, MA

Western D, Wright RM (1994) The background to community-based conservation. In: Western D, Wright RM (eds) (Strum SC (assoc. ed)) Natural connections: perspectives in community-based conservation. Island Press, Washington, DC, pp 1–12

Wheatley M (1999) Leadership and the new science. Berrett-Koehler, San Francisco, CA

Willer D, Lovagila MJ, Markovsky B (1997) Power and influence: a theoretical bridge. Soc Forces 76(2):571–603

Willey S, Perkins B (2005) Service lists Salt Creek Tiger Beetle as endangered. News release: U.S. Fish and Wildlife Service, 6 March 2007. http://mountain-prairie.fws.gov/pressrel/05-72.htm, accessed 10-20-2008

Wilson PF, Pearson RD (1995) Performance-based assessments: external, internal, and self-assessment tools for total quality management. ASQC Quality, Milwaukee, WI

Wilson RK (2006) Collaboration in context: rural change and community forestry in the Four Corners. Soc Nat Resour 19:53–70

Wollondeck JM, Yaffee SL (2000) Making collaboration work: lessons from a comprehensive assessment of over 200 wide ranging cases of collaboration in environmental management. Conserv Pract 1(1):17–25

Xiao C, McCright AM (2007) Environmental concern and sociodemographic variables: a study of statistical models. J Environ Educ 38(2):3–13

Yankelovich D (2001) The magic of dialogue: transforming conflict into cooperation. Touchstone, New York, NY

Zacharakis J, Morton LW, Rodecap J (2002) Citizen-led watershed projects: participatory research and environmental adult learning along Iowa's Maquoketa River. Adult Learn 13(2):19–23

Zacharakis J (2006) Conflict as a form of capital in controversial community development projects. J Ext 44(5):5FEA2. http://www.joe.org/joe/2006october/a2.shtml, accessed 8-22-2008

Index

A

Ad Hoc Task Force, 115
Agricultural management, 226–227
Agricultural pollution, 231
Agroecosystems Management Program
    (AMP), 200–202
Applegate Partnership, 49

B

Best management practices (BMP), 229

C

Causation and responsibility, 204–205, 210
CBRM. *See* Community-based resource
    management
Cheney Lake Watershed Inc. (CLWI).
        See also Rural-urban watershed
        governance, Central Kansas
    human population, 125
    mutual benefits, 130–131
    reservoir, 126–127
    urban-rural partnership, 128–129
    water quality, 127
    water supply, 125–126
Citizen capacity building
    citizen involvement
        community organizations, 249–250
        cultural differences, 250
        documentation and evaluation,
            251–252
        leadership development, 255
        management targets, 252–253
        political boundaries, 249
        watershed management, 248–249
    community, 139–140

CWO, 139
    education and science literacy, 254
    group leaders, 139
    Heartland Regional Water Coordination
        Initiative, 247–248
    individual capacity, 139–140
    political capacity, 141
    watershed project, 253–254
Citizen effects measurement.
        See Citizen participation
Citizen involvement
    civil society, 15–16
    community organizations, 249–250
    cultural differences, 250
    democracy, 25
    documentation and evaluation,
        251–252
    intervention
        catalytic influence of local champions,
            25, 26
        land management, 26–27
        science and technology interventions, 26
        social and farm level outcomes, 25
        spillover effect, 27
    leadership development, 255
    management targets, 252–253
    political boundaries, 249
    "right" outcomes
        agricultural land use, public
            conversations, 19
        ecosystem services, 19
        mission-directed public agencies, 21
        public standards and enforcement, 20
        social pressure and internalization,
            23–24
        social theory, 19, 21–22
        US watershed communities, 18–19

Citizen involvement (*cont.*)
  social outcomes
    Flora's social control pyramid, 28
    GREM, 28
    personal internal beliefs-knowledge-civic
        structure interaction, 26, 27
    physical landscape change, 27
  water, public commons
    citizen *vs.* citizen counteractions, 17
    resource conservation and protection, 16
    watershed management, 17, 18
  water quality issues, sociological analysis,
      20–21
  watershed management, 248–249
Citizen participation
  agencies and nonprofit organizations, 84
  amount and quality, 83
  Arnstein's classic ladder, 84
  breadth and depth assessment
    adaptive management approach, 90, 91
    agricultural land water runoff quality,
        92–93
    management plan, 91
    nominal, consultative, and
        decision-making participation, 90
    participation goals, 90, 91
  conflict and turmoil types
    agriculture regulation, 63
    environmental conflict, 62
    family conflict, 61
    farmers and government conflicts, 62–63
    importance of, 64–65
    neighbors conflict, 61–62
    state and federal agencies, 63
    technical and regulatory agencies, 63–64
  conflict's political dimension, 57–58
  Maquoketa River Watershed
    Concentrated Animal Feeding Operation
        threshold, 60
    corn–soybean rotation, 59–60
    limestone formation, 60
    nonpoint source pollution, 59
    sediment and nutrients, 58
    Soil and Water Conservation Districts, 60
    steering committee, 61
  NPS mitigation, 83, 84
  stakeholders, 84–85
  success strategy, 65–66
  types
    consultative participation, 88–89
    co-optation, 86
    decision making, 89

    nominal participation, 86–87
    pre-determined activities, 87–88
    program participation, 87
    water quality improvement, 84
Citizens Management Committee (CMC).
    See Rural-urban watershed
    governance, Central Kansas
Clean Water Act, 3
CLWI. *See* Cheney Lake Watershed Inc.
CNT. *See* Corn stalk nitrate test
Coalition of Watershed Towns (CWT), 116–117
Community-based resource management
    (CBRM), 133–134
Community of interests. *See also* Watershed
    management
  boundary conditions, 112
  fairness, 112
  institutionalization, 117–118
  reference terms, 111
Community watershed organizations (CWO),
    Pennsylvania
  agency participation, 140–141
  capacity building
    community, 139–140
    group leaders, 139
    individual capacity, 139–140
    political capacity, 141
  CBRM, 133–134
  definition, 135
  distribution, 137
  ecosystem health, 136, 138
  environmental education, 138–139
  impacts, 143–144
  watershed groups, 134–135
Community watershed planning, Vandalia
  agriculture *vs.* urban interests, 150
  chief operator relationship, 150–151
  community and technical assistance, 149
  design and implementation, 151–152
  development process, 153–154
  drinking water, 145
  EPA, 145
  goals and solutions, 150
  impaired waterbody, 146
  impaired water classification, 148
  partnership and characterization, 148–149
  progress and adjustments, 153
  reservoir, 147–148
  TMDL elements, 146–147
Conservation Districts of Iowa (CDI), 70
Conservation language
  citizen participants, 69

culture
    agriculture economics, 75
    farmer decision-making process, 75
    ILF, 78–79
    yields and profitability, 75, 76
discourse, 68–69
good farmer, 69–70
green language bandwagon, 78
language, powerful tool, 78
listening sessions
    conservation management systems, 71
    farming government, 71–72
    flooding, 74
    IDNR, 70–71
    ILF, 70, 80
    ISUE, 80
    management practices, 75
    soil erosion, 74
    water quality problem, 71, 73
motivation, 76–77
NRCS, 67, 68
re-languaging, 67, 68
resilient agricultural system, 77
Conservation management, 246
Conservation movement, 97
Conservation Reserve Program (CRP), 22, 225
Consultative participation, 88–89
Contemporary environmental movement, 97
Continuous improvement, 182–183
Co-optation, 86
Corn stalk nitrate test (CNT), 240–241
Correlations and causal model
    environmental attitude variable, 106–107
    hypothesis, 105–106
    pro-ecocentric environment, 106
Cross-cultural collaboration, Riparian restoration
    fieldwork
        tree plantation and cedar revetments,
            174–175
        tribal and nontribal members, 175–176
        Kansas tribes, 172–173
        legislature, 176
        partnership building, 176–178
        PBPN, 172
        Potawatomi reservation, 178–179
        red elm, 178
        sovereignty, 171
        tribal council, 173–174
CRP. See Conservation Reserve Program
Culture of resistance, 113
CWO. See Community watershed
        organizations, Pennsylvania

CWT. See Coalition of Watershed Towns

D
Data monitoring
    cost effectiveness, 234, 236– 237
    crop and livestock management, 234
Decision making participation, 89
Decision support systems (DSS), 218

E
Eastern Nebraska Saline Wetlands
    disappearance, 157–159
    emergent interview themes, 162–163
    interagency partnership, 168–169
    interviews and surveys, 161–162
    location, 169–170
    methods and data, 160–161
    perception differences
        agencies and voluntary organizations,
            231–233
        ecological protection and endangered
            species, 164–165
        landowners, 165–168
        planners, 163–164
    SWCP, 159–160
Environmental attitudes, 103, 104
Environmental attitude variable model, 106–107
Environmental concern, 98–99
Environmental education, 138–139
Environmental management
    accessible data
        modeling, 234, 236–237
        monitoring, 233–235
        on-farm demonstration, 237–238
    agriculture, 230–231
    BMP, 229
    citizen-directed facilitation, 232
    ISUE, 229
    participatory learning, 230
    performance-based, 181–182
    technical assistance education
        Hewitt Creek accomplishment,
            245–246
        incentive program, 238–240
        neighbor-to-neighbor learning, 245
        on-farm incentives, 241–242
        performance measurement, 240–241, 244
        scenarios, 243–244
        watershed council, 242–243
    watershed council facilitation, 232–233

Environmental Protection Agency (EPA),
    3, 36, 100, 145
Expectancy-value model, 95–96

F
Farmer decision makers
  automatic management behavior, 220–221
  conservation
    agriculture, 213–214
    community, 221–222
    practices, 222–223
  facilitative learning, 215–216
  manure management, 219–220
  mental maps (*See* Mental maps)
  record-based management
    field and farm water interception,
      224, 225
    scale distribution, 223
    soil testing, 225
    tillage practices, 224–226
Farmer learning circle
  actions or solutions, 206–207, 210
  causation and responsibility, 204–205, 210
  grassroots participation, 197–198
  highly degraded agricultural watershed,
    199–200
  organization and process, 200–202, 209
  participation, 205–206, 210
  Sugar Creek
    partners process, 209–211
    watershed, 198–199
  values and goals, 207–208, 210–211
  water quality problems, 202–204, 210
  watershed language and assumptions,
    208–209
Flora's agroecosystem management model,
    21, 22

G
Grass-roots ecosystem management
  (GREM), 28
Grass roots environmental management,
    54, 97
Grassroots participation, 197–198

H
Heartland Regional Water Coordination
    Initiative, 247–248
HEL. *See* Highly erodible lands

Hewitt Creek watershed
  accomplishments, 245–246
  continuous process, 195–196
  impairment, 184
  model
    assessment, 187–188
    awareness, 185–186
    evaluation, 192–193
    goals and plans, 188–189
    performance, 190–192
    targets, 189–190
Highly erodible lands (HEL), 225
Hydrological unit code (HUC), 23, 48, 59

I
IDNR. *See* Iowa Department of Natural
    Resources
ILF. *See* Iowa Learning Farm
Impaired waterbody, 146
Interagency partnership, 168–169
Iowa Department of Agriculture and
    Land Stewardship, 63
Iowa Department of Natural Resources
    (IDNR), 20, 46–47, 63, 68, 184
Iowa Farm Bureau (IFB), 70
Iowa Learning Farm (ILF), 68, 70, 78–80
Iowa State University Extension (ISUE),
    68, 80, 229

K
Kansas tribes, 172–173
Kansas WRAPS program, 35

L
Leadership, watershed management
  conditions needed, 33
  context and situation, 32
  federal Farm Bill legislation, 34
  grass roots leadership, 29–30
  natural resource decision making, 33–34
  social relationships and influence, 31–32
  trigger points, 37–38
  trust and quality relationships
    community group development, 37
    EPA, 36
    Kansas WRAPS program, 35
    "us" *vs.* "them" division, 36
  vision
    financial incentive strategy, 34

Flora's model, 33
impaired waters, 34–35
personal beliefs and attitudes, 34
self-interests and champion, 30
WRAPS, 31
watershed restoration, 39
watershed-wide changes, 37
Leopold Center for Sustainable
Agriculture, 70

M

Maquoketa Headwaters project, 232–233
Maquoketa River Watershed
Concentrated Animal Feeding Operation
threshold, 60
corn–soybean rotation, 59–60
limestone formation, 60
nonpoint source pollution, 59
sediment and nutrients, 58
Soil and Water Conservation Districts, 60
steering committee, 61
Memorandum of Agreement (MOA), 109
Mental maps
definition, 216–217
DSS, 218
judgement survey, 218–219
scientist and specialist roles, 217
water quality, 217–218
Multi-state water issue survey
core survey items, 100–101
EPA, 100
random sampling, 99–100

N

National Water Quality Inventory, 5
Natural Resource Conservation Service
(NRCS), 46–47, 59, 67
Natural resource management, 155
New York City (NYC)
agreement, 112–113
culture of resistance, 113
land use, 113–114
water supply system, 110
NMP. *See* Nutrient management plan
Nominal participation, 86–87
Nonpoint source (NPS) pollution, 3
NRCS. *See* Natural Resource Conservation
Service
Nutrient management plan (NMP), 239–240
NYC. *See* New York City

P

PBPN. *See* Prairie Band Potawatomi Nation
Performance-based management
citizen engagement, 183–184
continuous improvement, 182–183
effects, 193–194
environmental, 181–182
Hewitt Creek watershed model
assessment, 187–188
awareness, 185–186
continuous process, 195–196
evaluation, 193
goals and plans, 188–189
impairment, 184
performance, 190–192
targets, 189–190
Performance measurement, 240–241
Policy network approach, 123–124
Prairie Band Potawatomi Nation (PBPN),
12, 172
Precipitation and runoff, 3
Pre-determined activities participation,
87–88
Preservation movement, 97
Pro-ecocentric environment, 106
Property rights, 156–157

R

Record-based management
field and farm water interception,
224, 225
scale distribution, 223
soil testing, 225
tillage practices, 224–226
Red elm, 178
Regional water quality and environment
beliefs and attitudes, 95–96
concerns, 98–99
multi-state water issue survey
core survey items, 100–101
EPA, 100
random sampling, 99–100
perceptions
correlations and causal model,
105–107
environmental attitudes, 103, 104
surface and ground water, 101–103
watering practices, 103–105
willingness to learn, 103, 105
social issue, 95
US environmental movements, 96–97

Rural–urban interface, 11
Rural-urban watershed governance,
        Central Kansas
    CLWI
        human population, 125
        mutual benefits, 130–131
        partnership, 128–129
        reservoir, 126–127
        water quality, 127
        water supply, 125–126
    history, 124–125
    participation, 121–122
    policy network approach, 123–124
    resource management, 122–123

S
Saline Wetlands Conservation Partnership
        (SWCP), 159–160
Salt Creek tiger beetle, 157, 159
Soil conditioning index (SCI), 13, 240–241
State-level water quality, 11
Sugar Creek partners process
    actions and solutions, 210
    causation and responsibility, 210
    organization and process, 209
    participation, 210
    problem identification and definition, 210
    values and goals, 210–211
SWCP. See Saline Wetlands Conservation
        Partnership (SWCP)

T
Technical assistance environmental education
    Hewitt Creek accomplishment, 245–246
    incentive program, 238–240
    neighbor-to-neighbor learning, 245
    on-farm incentives, 241–242
    performance measurement, 240–241
The Nature Conservancy (TNC), 156
Total maximum daily load (TMDL), 12,
        146–147
Tribal council, 173–174
Tribe sovereignty, 171

U
United States Department of Agriculture
        (USDA), 18, 72
US Bureau of Land Management, 49

USDA Natural Resources Conservation
        Service, 10
US environmental movements, 96–97
US Forest Service, 49
US water quality concerns
    citizen participation, 8–9
    nonpoint source pollution, 4–6
    place-based decision making, 7–8
    watershed management, 6–7
US watershed communities, outcomes,
        18–19

W
WAC. See Watershed Agricultural Council
Water, learning
    balancing act
        bureaucratic risks and costs, 53
        grassroots community, 53, 54
        neighbor-to-neighbor or referent power,
                54–55
        polarization, 54
        unequal power relations, 54
    farmers environmental
                management goal, 47
    IDNR, 46–47
    influence analysis, political
                accounting system
        Coplin and O'Leary political
                calculations, 53
        NPS management behaviors
                and actions, 51
        political and social connections, 51
        water quality issue, 52–53
        watershed group efforts, 52
    information exchange and
                social pressure, 46
    information power, 50
    neighbor-to-neighbor relationships, 48
    NRCS, 46–47
    organizational power and influence, 49–50
    peer-to-peer influence, 48
    political problem
        sources, influence and power, 42–43
        water pollution, 42
    power and influence
        coercive and reward power, 43–44
        expert power, 44
        informational power, 44–45
        legitimate power, 44
        referent power, 44

prestige and status, 45–46
survey and qualitative data, 41
Water pollution, 95
Water quality perceptions
  correlations and causal model,
    105–107
  environmental attitudes, 103, 104
  surface and ground water,
    101–103
  watering practices, 103–105
  willingness to learn, 103, 105
Watershed Agricultural Council
  (WAC), 114
Watershed management, 248–249
  community of interests
    boundary conditions, 112
    fairness, 112
    institutionalization, 117–118
    reference terms, 111

efforts
  Ad Hoc Task Force, 115
  CWT, 116–117
  identity, 114–115
  MOA, 109
  NYC
    agreement, 112–113
    culture of resistance, 113
    land use, 113–114
    water supply system, 110
Watershed protection
  Eastern Nebraska Saline Wetlands
    (*See* Eastern Nebraska Saline
    Wetlands)
  local conflict *vs.* property rights, 156–157
  natural resource management, 155
Watershed resource management, 122–123
Watershed Restoration and Protection Strategy
  (WRAPS), 31, 35